Selected Titles in This Series

43 **N. V. Krylov,** Introduction to the theory of random processes, 2002
42 **Jin Hong and Seok-Jin Kang,** Introduction to quantum groups and crystal bases, 2002
41 **Georgi V. Smirnov,** Introduction to the theory of differential inclusions, 2002
40 **Robert E. Greene and Steven G. Krantz,** Function theory of one complex variable, 2002
39 **Larry C. Grove,** Classical groups and geometric algebra, 2002
38 **Elton P. Hsu,** Stochastic analysis on manifolds, 2002
37 **Hershel M. Farkas and Irwin Kra,** Theta constants, Riemann surfaces and the modular group, 2001
36 **Martin Schechter,** Principles of functional analysis, second edition, 2002
35 **James F. Davis and Paul Kirk,** Lecture notes in algebraic topology, 2001
34 **Sigurdur Helgason,** Differential geometry, Lie groups, and symmetric spaces, 2001
33 **Dmitri Burago, Yuri Burago, and Sergei Ivanov,** A course in metric geometry, 2001
32 **Robert G. Bartle,** A modern theory of integration, 2001
31 **Ralf Korn and Elke Korn,** Option pricing and portfolio optimization: Modern methods of financial mathematics, 2001
30 **J. C. McConnell and J. C. Robson,** Noncommutative Noetherian rings, 2001
29 **Javier Duoandikoetxea,** Fourier analysis, 2001
28 **Liviu I. Nicolaescu,** Notes on Seiberg-Witten theory, 2000
27 **Thierry Aubin,** A course in differential geometry, 2001
26 **Rolf Berndt,** An introduction to symplectic geometry, 2001
25 **Thomas Friedrich,** Dirac operators in Riemannian geometry, 2000
24 **Helmut Koch,** Number theory: Algebraic numbers and functions, 2000
23 **Alberto Candel and Lawrence Conlon,** Foliations I, 2000
22 **Günter R. Krause and Thomas H. Lenagan,** Growth of algebras and Gelfand-Kirillov dimension, 2000
21 **John B. Conway,** A course in operator theory, 2000
20 **Robert E. Gompf and András I. Stipsicz,** 4-manifolds and Kirby calculus, 1999
19 **Lawrence C. Evans,** Partial differential equations, 1998
18 **Winfried Just and Martin Weese,** Discovering modern set theory. II: Set-theoretic tools for every mathematician, 1997
17 **Henryk Iwaniec,** Topics in classical automorphic forms, 1997
16 **Richard V. Kadison and John R. Ringrose,** Fundamentals of the theory of operator algebras. Volume II: Advanced theory, 1997
15 **Richard V. Kadison and John R. Ringrose,** Fundamentals of the theory of operator algebras. Volume I: Elementary theory, 1997
14 **Elliott H. Lieb and Michael Loss,** Analysis, 1997
13 **Paul C. Shields,** The ergodic theory of discrete sample paths, 1996
12 **N. V. Krylov,** Lectures on elliptic and parabolic equations in Hölder spaces, 1996
11 **Jacques Dixmier,** Enveloping algebras, 1996 Printing
10 **Barry Simon,** Representations of finite and compact groups, 1996
9 **Dino Lorenzini,** An invitation to arithmetic geometry, 1996
8 **Winfried Just and Martin Weese,** Discovering modern set theory. I: The basics, 1996
7 **Gerald J. Janusz,** Algebraic number fields, second edition, 1996
6 **Jens Carsten Jantzen,** Lectures on quantum groups, 1996
5 **Rick Miranda,** Algebraic curves and Riemann surfaces, 1995
4 **Russell A. Gordon,** The integrals of Lebesgue, Denjoy, Perron, and Henstock, 1994

(Continued in the back of this publication)

Introduction to the Theory of Random Processes

Introduction to the Theory of Random Processes

N. V. Krylov

Graduate Studies
in Mathematics
Volume 43

American Mathematical Society
Providence, Rhode Island

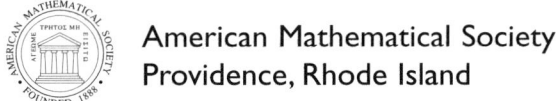

2000 *Mathematics Subject Classification.* Primary 60-01; Secondary 60G99.

The author was supported in part by NSF Grant DMS-9876586

ABSTRACT. These lecture notes concentrate on some general facts and ideas of the theory of stochastic processes. The main objects of study are the Wiener processes, the stationary processes, the infinitely divisible processes, and the Itô stochastic equations.

Although it is not possible to cover even a noticeable portion of the topics listed above in a short course, the author sincerely hopes that after having followed the material presented here the reader will have acquired a good understanding of what kind of results are available and what kind of techniques are used to obtain them.

These notes are intended for graduate students and scientists in mathematics, physics and engineering interested in the theory of random processes and its applications.

Library of Congress Cataloging-in-Publication Data

Krylov, N. V. (Nikolaĭ Vladimirovich)
Introduction to the theory of random processes / N. V. Krylov
p. cm. – (Graduate studies in mathematics, ISSN 1065-7339; v. 43)
Includes bibliographical references and index.
ISBN 0-8218-2985-8 (alk. paper)
1. Stochastic processes. I. Title. II. Series.
QA274.K79 2002
519.2/3–dc21 2002018241

♾ The paper used in this book is acid-free and falls within the guidelines
established to ensure permanence and durability.
Visit the AMS home page at URL: http://www.ams.org/

10 9 8 7 6 5 4 3 2 1 07 06 05 04 03 02

Contents

Preface

For about ten years between 1973 and 1986 the author was delivering a one-year topics course "Random Processes" at the Department of Mechanics and Mathematics of Moscow State University. This topics course was obligatory for third-fourth year undergraduate students (about 20 years of age) with major in probability theory and its applications. With great sympathy I remember my first students in this course: M. Safonov, A. Veretennikov, S. Anulova, and L. Mikhailovskaya. During these years the contents of the course gradually evolved, simplifying and shortening to the shape which has been presented in two 83 and 73 page long rotaprint lecture notes published by Moscow State University in 1986 and 1987. In 1990 I emigrated to the USA and in 1998 got the opportunity to present parts of the same course as a one-quarter topics course in probability theory for graduate students at the University of Minnesota. I thus had the opportunity to test the course in the USA as well as on several generations of students in Russia. What the reader finds below is a somewhat extended version of my lectures and the recitations which went along with the lectures in Russia.

The theory of random processes is an extremely vast branch of mathematics which cannot be covered even in ten one-year topics courses with minimal intersection of contents. Therefore, the intent of this book is to get the reader acquainted only with some parts of the theory. The choice of these parts was mainly defined by the duration of the course and the author's taste and interests. However, there is no doubt that the ideas, facts, and techniques presented here will be useful if the reader decides to move on and study some other parts of the theory of random processes.

From the table of contents the reader can see that the main topics of the book are the Wiener process, stationary processes, infinitely divisible processes, and Itô integral and stochastic equations. Chapters 1 and 3 are devoted to some techniques needed in other chapters. In Chapter 1 we discuss some general facts from probability theory and stochastic processes from the point of view of probability measures on Polish spaces. The results of this chapter help construct the Wiener process by using Donsker's invariance principle. They also play an important role in other issues, for instance, in statistics of random processes. In Chapter 3 we present basics of discrete time martingales, which then are used in one way or another in all subsequent chapters. Another common feature of all chapters excluding Chapter 1 is that we use stochastic integration with respect to random orthogonal measures. In particular, we use it for spectral representation of trajectories of stationary processes and for proving that Gaussian stationary processes with rational spectral densities are components of solutions to stochastic equations. In the case of infinitely divisible processes, stochastic integration allows us to obtain a representation of trajectories through jump measures. Apart from this and from the obvious connection between the Wiener process and Itô's calculus, all other chapters are independent and can be read in any order.

The book is designed as a textbook. Therefore it does not contain any new theoretical material but rather a new compilation of some known facts, methods and ways of presenting the material. A relative novelty in Chapter 2 is viewing the Itô stochastic integral as a particular case of the integral of nonrandom functions against random orthogonal measures. In Chapter 6 we give two proofs of Itô's formula: one is more or less traditional and the other is based on using stochastic intervals. There are about 128 exercises in the book. About 41 of them are used in the main text and are marked with an asterisk. The bibliography contains some references we use in the lectures and which can also be recommended as a source of additional reading on the subjects presented here, deeper results, and further references.

The author is sincerely grateful to Wonjae Chang, Kyeong-Hun Kim, and Kijung Lee, who read parts of the book and pointed out many errors, to Dan Stroock for his friendly critisizm of the first draft, and to Naresh Jain for useful suggestions.

Nicolai Krylov
Minneapolis, January 2001

Chapter 1

Generalities

This chapter is of an introductory nature. We start with recalling some basic probabilistic notions and facts in Sec. 1. Actually, the reader is supposed to be familiar with the material of this rather short section, which in no way is intended to be a systematic introduction to probability theory. All missing details can be found, for instance, in excellent books by R. Dudley [**Du**] and D. Stroock [**St**]. In Sec. 2 we discuss measures on Polish spaces. Quite often this subject is also included in courses on probability theory. Sec. 3 is devoted to the notion of random process, and in Sec. 4 we discuss the relation between continuous random processes and measures on the space of continuous functions.

1. Some selected topics from probability theory

The purpose of this section is to remember some familiar tunes and get warmed up. We just want to refresh our memory, recall some standard notions and facts, and introduce the notation to be used in the future.

Let Ω be a set and $\mathcal{F}$ a collection of its subsets.

1. Definition. We say that $\mathcal{F}$ is a *σ-field* if

(i) $\Omega \in \mathcal{F}$,

(ii) for every $A_1, ..., A_n, ...$ such that $A_n \in \mathcal{F}$, we have $\bigcup_n A_n \in \mathcal{F}$,

(iii) if $A \in \mathcal{F}$, then $A^c := \Omega \setminus A \in \mathcal{F}$.

In the case when $\mathcal{F}$ is a σ-field the couple $(\Omega, \mathcal{F})$ is called *a measurable space*, and elements of $\mathcal{F}$ are called *events*.

2. Example. Let Ω be a set. Then $\mathcal{F} := \{\emptyset, \Omega\}$ is a σ-field which is called the trivial σ-field.

3. Example. Let Ω be a set. Then the family Σ of all its subsets is a σ-field.

Example 3 shows, in particular, that if $\mathcal{F}$ is a family of subsets of Ω, then there always exists at least one σ-field containing $\mathcal{F}$: $\mathcal{F} \subset \Sigma$. Furthermore, it is easy to understand that, given a collection of σ-fields $\mathcal{F}^\alpha$ of subsets of Ω, where α runs through a set of indices, the set of all subsets of Ω each of which belongs to every σ-field $\mathcal{F}^\alpha$ is again a σ-field. In other words the intersection of every nonempty collection of σ-fields is a σ-field. In view of Example 3, it makes sense to consider the intersection of all σ-fields containing a given family $\mathcal{F}$ of subsets of Ω, and this intersection is a σ-field. Hence the smallest σ-field containing $\mathcal{F}$ exists. It is called the σ-field *generated by* $\mathcal{F}$ and is denoted by $\sigma(\mathcal{F})$.

If X is a closed subset of $\mathbb{R}^d$, the σ-field of its subsets generated by the collection of intersections of all closed balls in $\mathbb{R}^d$ with X is called the Borel σ-field and is denoted by $\mathfrak{B}(X)$. Elements of $\mathfrak{B}(X)$ are called Borel subsets of X.

Assume that $\mathcal{F}$ is a σ-field (then of course $\sigma(\mathcal{F}) = \mathcal{F}$). Suppose that to every $A \in \mathcal{F}$ there is assigned a number $P(A)$.

4. Definition. We say that $P(\cdot)$ is a *probability measure on* $(\Omega, \mathcal{F})$ or on $\mathcal{F}$ if

(i) $P(A) \geq 0$ and $P(\Omega) = 1$,

(ii) for every sequence of pairwise disjoint $A_1, ..., A_n, ... \in \mathcal{F}$, we have

$$P(\bigcup_n A_n) = \sum_n P(A_n).$$

If on a measurable space $(\Omega, \mathcal{F})$ there is defined a probability measure P, the triple $(\Omega, \mathcal{F}, P)$ is called *a probability space*.

5. Example. The triple, consisting of $[0,1]$ $(= \Omega)$, the σ-field $\mathfrak{B}([0,1])$ of Borel subsets of $[0,1]$ (taken as $\mathcal{F}$) and Lebesgue measure ℓ (as P) is a probability space.

Let $(\Omega, \mathcal{F}, P)$ be a probability space and $A \subset \Omega$ (not necessarily $A \in \mathcal{F}$).

6. Definition. We say that A has zero probability and write $P(A) = 0$ if there exists a set $B \in \mathcal{F}$ such that $A \subset B$ and $P(B) = 0$. The family of all subsets of Ω of type $C \cup A$, where $C \in \mathcal{F}$ and A has zero probability, is denoted by $\mathcal{F}^P$ and called *the completion of* $\mathcal{F}$ *with respect to* P. If $\mathcal{G} \subset \mathcal{F}$ is a sub-σ-field of $\mathcal{F}$, one completes G in the same way by using again events of zero probability (from $(\Omega, \mathcal{F}, P)$ but not $(\Omega, \mathcal{G}, P)$).

7. Exercise*. Prove that $\mathcal{F}^P$ is a σ-field.

The measure P extends to $\mathcal{F}^P$ by the formula $P(C \cup A) = P(C)$ if $C \in \mathcal{F}$ and $P(A) = 0$. It is easy to prove that this extension is well defined, preserves the values of P on $\mathcal{F}$ and yields a probability measure on $\mathcal{F}^P$.

8. Definition. The σ-field $\mathcal{F}$ is said to be *complete* (with respect to P) if $\mathcal{F}^P = \mathcal{F}$. The probability space $(\Omega, \mathcal{F}, P)$ is said to be complete if $\mathcal{F}^P = \mathcal{F}$, that is, if $\mathcal{F}$ contains all sets of zero probability. If $\mathcal{G} \subset \mathcal{F}$ is a sub-σ-field of $\mathcal{F}$ containing all sets of zero probability, it is also called complete.

The above argument shows that every probability space $(\Omega, \mathcal{F}, P)$ admits a completion $(\Omega, \mathcal{F}^P, P)$. In general there are probability spaces which are not complete. In particular, in Example 5 the completion of $\mathfrak{B}([0,1])$ with respect to ℓ is *the σ-field of Lebesgue sets* (or *Lebesgue σ-field*), which does not coincide with $\mathfrak{B}([0,1])$. In other words, there are sets of measure zero which are not Borel.

9. Exercise. Let f be the Cantor function on $[0,1]$, and let C be a non-Borel subset of $[0,1] \setminus \rho$, where ρ is the set of all rational numbers. Existence of such C is guaranteed, for instance, by Vitali's example. Prove that $\{x : f(x) \in C\}$ has Lebesgue measure zero and is not Borel.

By definition, for every $B \in \mathcal{F}^P$, there exists $C \in \mathcal{F}$ such that $P(B \setminus C) = 0$. Therefore, the advantages of considering $\mathcal{F}^P$ may look very slim. However, sometimes it turns out to be very convenient to pass to $\mathcal{F}^P$, because then more sets become measurable and tractable in the framework of measure theory. It is worth noting the following important result even though it will not be used in the future. It turns out that the projection on the x-axis of a Borel subset of $\mathbb{R}^2$ is not necessarily Borel, but is always a Lebesgue set (see, for instance, [**Me**]). Therefore, if $f(x,y)$ is a Borel function on $\mathbb{R}^2$, then for the function $\bar{f}(x) := \sup\{f(x,y) : y \in \mathbb{R}\}$ and every $c \in \mathbb{R}$ we have

$$\{x : \bar{f}(x) > c\} = \{x : \exists y \quad \text{such that} \quad f(x,y) > c\} \in \mathfrak{B}^{\ell}(\mathbb{R}).$$

It follows that $\bar{f}$ is Lebesgue measurable (but not necessarily Borel measurable) and it makes sense to consider its integral against dx. On the other hand, one knows that for every $\mathcal{F}^P$-measurable function there exists an $\mathcal{F}$-measurable one equal to the original *almost surely*, that is, such that the set where they are different has zero probability. It follows that there exists a Borel function equal to $\bar{f}(x)$ almost everywhere. However the last sentence is just a long way of saying that $\bar{f}(x)$ is measurable, and it also calls for new notation for the modification, which can make exposition quite cumbersome.

10. Lemma. *Let Ω and X be sets and let ξ be a function defined on Ω with values in X. For every $B \subset X$ set $\xi^{-1}(B) = \{\omega \in \Omega : \xi(\omega) \in X\}$. Then*

(i) ξ^{-1} *as a mapping between sets preserves all set-theoretic operations (for instance, if we are given a family of subsets* B^α *of* X *indexed by* α, *then* $\xi^{-1}(\bigcup_\alpha B^\alpha) = \bigcup_\alpha \xi^{-1}(B^\alpha)$, *and so on),*

(ii) *if* $\mathcal{F}$ *is a* σ*-field of subsets of* Ω, *then*

$$\{B : B \subset X, \xi^{-1}(B) \in \mathcal{F}\}$$

is a σ*-field of subsets of* X.

We leave the proof of these simple facts to the reader.

If $\xi : \Omega \to X$ and there is a σ-field $\mathfrak{B}$ of subsets of X, we denote $\sigma(\xi) := \xi^{-1}(\mathfrak{B}) := \{\xi^{-1}(B) : B \in \mathfrak{B}\}$. By Lemma 10 (i) the family $\xi^{-1}(\mathfrak{B})$ is a σ-field. It is called *the* σ*-field generated by* ξ. Observe that, by definition, each element of $\sigma(\xi)$ is representable as $\{\omega : \xi(\omega) \in B\}$ for some $B \in \mathfrak{B}$.

11. Definition. Let $(\Omega, \mathcal{F})$ and $(X, \mathfrak{B})$ be measurable spaces, and let $\xi : \Omega \to X$ be a function. We say that ξ is a *random variable* if $\sigma(\xi) \subset \mathcal{F}$.

If, in addition, $(\Omega, \mathcal{F}, P)$ is a probability space and ξ is a random variable, the function defined on $\mathfrak{B}$ by the formula

$$P\xi^{-1}(B) = P(\xi^{-1}(B)) = P\{\omega : \xi(\omega) \in B\}$$

is called *the distribution of* ξ. By Lemma 10 (i) the function $P\xi^{-1}$ is a probability measure on $\mathfrak{B}$. One also uses the notation

$$F_\xi = P\xi^{-1}.$$

It turns out that every probability measure is the distribution of a random variable.

12. Theorem. *Let* μ *be a probability measure on a measurable space* $(X, \mathfrak{B})$. *Then there exist a probability space* $(\Omega, \mathcal{F}, P)$ *and an* X*-valued random variable defined on this space such that* $F_\xi = \mu$.

Proof. Let $(\Omega, \mathcal{F}, P) = (X, \mathfrak{B}, \mu)$ and $\xi(x) = x$. Then $\{x : \xi(x) \in B\} = B$. Hence for every $B \in \mathfrak{B}$ we have $F_\xi(B) = \mu(B)$, and the theorem is proved.

Remember that if ξ is a real-valued random variable defined on a probability space $(\Omega, \mathcal{F}, P)$ and at least one of the integrals

$$\int_\Omega \xi_+(\omega)\, P(d\omega), \quad \int_\Omega \xi_-(\omega)\, P(d\omega)$$

($\xi_\pm := (|\xi| \pm \xi)/2$) is finite, then by *the expectation of* ξ we mean

$$E\xi := \int_\Omega \xi(\omega)\,P(d\omega) := \int_\Omega \xi_+(\omega)\,P(d\omega) - \int_\Omega \xi_-(\omega)\,P(d\omega).$$

The next theorem relates expectations to distributions.

13. Theorem. *Let $(\Omega, \mathcal{F}, P)$ be a probability space, $(X, \mathfrak{B})$ a measurable space and $\xi : \Omega \to X$ a random variable. Let f be a measurable mapping from $(X, \mathfrak{B})$ to $([0, \infty), \mathfrak{B}[0, \infty))$. Then $f(\xi)$ is a random variable and*

$$Ef(\xi) = \int_X f(x)\,F_\xi(dx). \tag{1}$$

Proof. For $t \geq 0$, let $[t]$ be the integer part of t and $\kappa_n(t) = 2^{-n}[2^n t]$. Drawing the graph of κ_n makes it clear that $0 \leq t - \kappa_n(t) \leq 2^{-n}$, κ_n increases when n increases, and κ_n are Borel functions. Furthermore, the variables $f(\xi)$, $\kappa_n(f(\xi))$, $\kappa_n(f(x))$ are appropriately measurable and, by the monotone convergence theorem,

$$Ef(\xi) = \lim_{n\to\infty} E\kappa_n(f(\xi)), \quad \int_X f(x)\,F_\xi(dx) = \lim_{n\to\infty} \int_X \kappa_n(f(x))\,F_\xi(dx).$$

It follows that it suffices to prove the theorem for functions $\kappa_n(f)$. Each of them is measurable and only takes countably many nonnegative values; that is, it has the form

$$\sum_k c_k I_{B_k}(x),$$

where $B_k \in \mathfrak{B}$ and $c_k \geq 0$. It only remains to notice that by definition

$$EI_{B_k}(\xi) = P\{\xi \in B_k\} = F_\xi(B_k) = \int_X I_{B_k}(x)\,F_\xi(dx)$$

and by the monotone convergence theorem

$$E\sum_k c_k I_{B_k}(\xi) = \sum_k c_k E I_{B_k}(\xi) = \int_X \sum_k c_k I_{B_k}(x)\,F_\xi(dx).$$

The theorem is proved.

Notice that (1) also holds for f taking values of different signs whenever at least one side of (1) makes sense. This follows easily from the equality $f = f_+ - f_-$ and from (1) applied to $f_\pm$.

2. Some facts from measure theory on Polish spaces

In this book the only Polish spaces we will be dealing with are Euclidean spaces and the space of continuous functions defined on $[0, 1]$.

2:1. Definitions and simple facts. A complete separable metric space is called a *Polish space*. Let X be a Polish space with metric $\rho(x,y)$. By definition the closed ball of radius r centered at x is

$$B_r(x) = \{y : \rho(x,y) \le r\}.$$

The smallest σ-field of subsets of X containing all closed balls is called *the Borel σ-field* and is denoted $\mathfrak{B}(X)$. Elements of $\mathfrak{B}(X)$ are called *Borel sets*.

The structure of an arbitrary Borel set, even in $\mathbb{R}$, is extremely complex. However, very often working with *all* Borel sets is rather convenient.

Observe that

$$\{y : \rho(x,y) < r\} = \bigcup_n \{y : \rho(x,y) \le r - 1/n\}.$$

Therefore, open balls are Borel. Furthermore, since X is separable, each open set can be represented as the countable union of certain open balls. Therefore, open sets are Borel. Their complements, which are arbitrary closed sets, are Borel sets as well. By the way, it follows from this discussion that one could equivalently define the Borel σ-field as the smallest σ-field of subsets of X containing all open balls.

If X and Y are Polish spaces, and $f : X \to Y$, then the function f is called *a Borel function* if

$$f^{-1}(B) := \{x : f(x) \in B\} \in \mathfrak{B}(X) \quad \forall B \in \mathfrak{B}(Y).$$

In other words f is a Borel function if $f : X \to Y$ is a random variable with respect to the σ-fields $\mathfrak{B}(X)$ and $\mathfrak{B}(Y)$. An example of Borel functions is given in the following theorem.

1. Theorem. *Let X and Y be Polish spaces, and let $f : X \to Y$ be a continuous function. Then f is Borel.*

Proof. Remember that by Lemma 1.10 the collection

$$\Sigma := \{B \subset Y : f^{-1}(B) \in \mathfrak{B}(X)\}$$

is a σ-field. Next, for every $B_r(y) \subset Y$ the set $f^{-1}(B_r(y))$ is closed because of the continuity of f. Hence $B_r(x) \in \Sigma$. Since $\mathfrak{B}(Y)$ is the smallest σ-field containing all $B_r(x)$, we have $\mathfrak{B}(Y) \subset \Sigma$, which is the same as saying that f is Borel. The theorem is proved.

Let us emphasize a very important feature of the above proof. Instead of taking a particular $B \in \mathfrak{B}(Y)$ and proving that $f^{-1}(B) \in \mathfrak{B}(X)$, we took the collection of *all* sets possessing a desired property. This device will be used quite often.

Next, we are going to treat measures on Polish spaces. We recall that a measure is called finite if all its values belong to $(-\infty, \infty)$. Actually, it is safe to say that everywhere in the book we are always dealing with *nonnegative* measures. The only exception is encountered in Remark 17, and even there we could avoid using signed measures if we rely on π- and λ-systems, which come somewhat later in Sec. 2.3.

2. Theorem. *Let X be a Polish space and μ a finite nonnegative measure on $(X, \mathfrak{B}(X))$. Then μ is regular in the sense that for every $B \in \mathfrak{B}(X)$ and $\varepsilon > 0$ there exist an open set G and a closed set Γ satisfying*

$$G \supset B \supset \Gamma, \quad \mu(G \setminus \Gamma) \leq \varepsilon. \tag{1}$$

Proof. Take a finite nonnegative measure μ on $(X, \mathfrak{B}(X))$ and call a set $B \in \mathfrak{B}(X)$ *regular* if for every $\varepsilon > 0$ there exist open G and closed Γ satisfying (1).

Let Σ be the set of all "regular" sets. We are going to prove that

(i) Σ is a σ-field, and

(ii) $B_r(x) \in \Sigma$.

Then by the definition of $\mathfrak{B}(X)$ we have $\mathfrak{B}(X) \subset \Sigma$, and this is exactly what we need.

Statement (ii) is almost trivial since, for every $n \geq 1$,

$$\Gamma := B_r(x) \subset B_r(x) \subset \{x : \rho(x, y) < r + 1/n\} =: G_n,$$

where Γ is closed, the G_n are open and $\mu(G_n \setminus \Gamma) \to 0$ since the sets $G_n \setminus \Gamma$ are nested and their intersection is empty.

To prove (i), first notice that $X \in \Sigma$ as a set open and closed simultaneously. Furthermore, the complement of an open (closed) set is a closed (respectively, open) set and if $G \supset B \supset \Gamma$, then $\Gamma^c \supset B^c \supset G^c$ with

$$\Gamma^c \setminus G^c = G \setminus \Gamma.$$

This shows that if $B \in \Sigma$, then $B^c \in \Sigma$. It only remains to check that countable unions of elements of Σ belong to Σ.

Let $B_n \in \Sigma$, $n = 1, 2, 3, \ldots$, $\varepsilon > 0$, and let G_n be open and Γ_n be closed and such that

$$G_n \supset B_n \supset \Gamma_n, \quad \mu(G_n \setminus \Gamma_n) \leq \varepsilon 2^{-n}.$$

Define

$$B=\bigcup_n B_n,\quad G=\bigcup_n G_n,\quad D_n=\bigcup_{i=1}^n \Gamma_i.$$

Then G is open, D_n is closed, and obviously $G\setminus D_n$ are nested, so that

$$\lim_{n\to\infty}\mu(G\setminus D_n)=\mu(G\setminus D_\infty)\le\sum_n \mu(G_n\setminus\Gamma_n)\le\varepsilon.$$

Hence, for appropriate n we have $\mu(G\setminus D_n)\le 2\varepsilon$, and this brings the proof to an end.

3. Corollary. *If μ_1 and μ_2 are finite nonnegative measures on $(X,\mathfrak{B}(X))$ and $\mu_1(\Gamma)=\mu_2(\Gamma)$ for all closed Γ, then $\mu_1=\mu_2$.*

Indeed, then $\mu_1(X)=\mu_2(X)$ (X is closed) and hence the μ_i's also coincide on all open subsets of X. But then they coincide on all Borel sets, as seen from

$$\mu_i(G_i)\ge\mu_i(B)\ge\mu_i(\Gamma_i),\quad \mu_i(G\setminus\Gamma)\le\varepsilon,$$

where $G=G_1\cap G_2$, $\Gamma=\Gamma_1\cup\Gamma_2$ and $G\setminus\Gamma$ is open.

4. Theorem. *If μ_1 and μ_2 are finite nonnegative measures on $(X,\mathfrak{B}(X))$ and*

$$\int_X f(x)\,\mu_1(dx)=\int_X f(x)\,\mu_2(dx)$$

for every bounded continuous f, then $\mu_1=\mu_2$.

Proof. By the preceding corollary we only need to check that $\mu_1=\mu_2$ on closed sets. Take a closed set Γ and let

$$\rho(x,\Gamma)=\inf\{\rho(x,y):y\in\Gamma\}.$$

Since the absolute value of a difference of inf's is not greater than the sup of the absolute values of the differences and since $|\rho(x,y)-\rho(z,y)|\le\rho(x,z)$, we have that $|\rho(x,\Gamma)-\rho(z,\Gamma)|\le\rho(x,z)$, which implies that $\rho(x,\Gamma)$ is continuous. Furthermore,

$$\rho(x,\Gamma)>0\iff x\notin\Gamma$$

since Γ is closed. Hence, for the continuous function

$$f_n(x):=(1+n\rho(x,\Gamma))^{-1}$$

we have $1\ge f_n(x)\downarrow I_\Gamma(x)$, so that by the dominated convergence theorem

$$\mu_1(\Gamma)=\int I_\Gamma\,\mu_1(dx)=\lim_{n\to\infty}\int f_n\,\mu_1(dx)=\lim_{n\to\infty}\int f_n\,\mu_2(dx)=\mu_2(\Gamma).$$

The theorem is proved.

2:2. Tightness and convergence of measures. As we have mentioned in the Preface, the results of this chapter help construct the Wiener process by using a version of the central limit theorem for random walks known as Donsker's invariance principle. Therefore we turn our attention to studying convergence of measures on Polish spaces. An important property of a measure on a Polish space is its tightness, which is expressed in the following terms.

5. Theorem (Ulam). *Let μ be a finite nonnegative measure on $(X, \mathfrak{B}(X))$. Then for every $\varepsilon > 0$ there exists a compact set $K \subset X$ such that $\mu(K^c) \leq \varepsilon$.*

Proof. Let $\{x_i : i = 1, 2, 3, ...\}$ be a dense subset of X. Observe that for every $n \geq 1$

$$\bigcup_i B_{1/n}(x_i) = X.$$

Therefore, there exists an i_n such that

$$\mu(\bigcup_{i \leq i_n} B_{1/n}(x_i)) \geq \mu(X) - \varepsilon 2^{-n}. \tag{2}$$

Now define

$$K = \bigcap_{n \geq 1} \bigcup_{i \leq i_n} B_{1/n}(x_i). \tag{3}$$

Observe that K is totally bounded in the sense that, for every $\varepsilon > 0$, there exists a finite set $A = \{x_1, ..., x_{i(\varepsilon)}\}$, called an ε-net, such that every point of K is in the ε-neighborhood of at least one point in A. Indeed, it suffices to take $i(\varepsilon) = i_n$ with any $n \geq 1/\varepsilon$.

In addition $\bigcup_{i \leq i_n} B_{1/n}(x_i)$ is closed as a finite union of closed sets, and then K is closed as the intersection of closed sets. It follows that K is a compact set (see Exercise 6). Now it only remains to notice that

$$\mu(K^c) \leq \sum_n \mu((\bigcup_{i \leq i_n} B_{1/n}(x_i))^c) \leq \sum_n \varepsilon 2^{-n}.$$

The theorem is proved.

6. Exercise*. Prove that the following are equivalent:

(i) K is a totally bounded closed set.

(ii) For every sequence of points $x_n \in K$, there is a subsequence $x_{n'}$ which converges to an element of K.

7. Corollary. *For every Borel B and $\varepsilon > 0$ there exists a compact set $\Gamma \subset B$ such that $\mu(B \setminus \Gamma) \leq \varepsilon$.*

Now we consider the issue of convergence of measures on X.

8. Definition. Let μ and μ_n be finite nonnegative measures on $(X, \mathfrak{B}(X))$. We say that μ_n *converge weakly to* μ and write $\mu_n \xrightarrow{w} \mu$ if for every bounded continuous function f

$$\int_X f\, \mu_n(dx) \to \int_X f\, \mu(dx) \tag{4}$$

A family $\mathcal{M}$ of finite measures on $(X, \mathfrak{B}(X))$ is called *relatively weakly (sequentially) compact* if every sequence of elements of $\mathcal{M}$ has a weakly convergent subsequence.

9. Exercise*. Let ξ, ξ_n be random variables with values in X defined on some probability spaces. Assume that the distributions of ξ_n on $(X, \mathfrak{B}(X))$ converge weakly to the distribution of ξ. Let $f(x)$ be a real-valued continuous function on X. Prove that the distributions of $f(\xi_n)$ converge weakly to the distribution of $f(\xi)$.

10. Exercise*. Let $\mathcal{M} = \{\mu_1, \mu_2, ...\}$ be a sequence of nonnegative finite measures on $(X, \mathfrak{B}(X))$ and let μ be a nonnegative measure on $(X, \mathfrak{B}(X))$. Prove that if every sequence of elements of $\mathcal{M}$ has a subsequence weakly convergent to μ, then $\mu_n \xrightarrow{w} \mu$.

11. Theorem. *Let μ, $\mu_n, n = 1, 2, 3, ...,$ be nonnegative finite measures on $(X, \mathfrak{B}(X))$. Then the following conditions are equivalent:*

(i) $\mu_n \xrightarrow{w} \mu$,

(ii) $\mu(\Gamma) \geq \overline{\lim_{n\to\infty}}\, \mu_n(\Gamma)$ *for every closed* Γ *and* $\mu(X) = \lim_{n\to\infty} \mu_n(X)$,

(iii) $\mu(G) \leq \underline{\lim}_{n\to\infty}\, \mu_n(G)$ *for every open* G *and* $\mu(X) = \lim_{n\to\infty} \mu_n(X)$,

(iv) $\mu(B) = \lim_{n\to\infty} \mu_n(B)$ *for every Borel* B *such that* $\mu(\partial B) = 0$,

(v) $\int f\, \mu_n(dx) \to \int f\, \mu(dx)$ *for every Borel bounded* f *such that* $\mu(\Delta_f) = 0$, *where* Δ_f *is the set of all points at which* f *is discontinuous.*

Proof. (i) $\Longrightarrow$(ii). Take a closed set Γ and define f_n as in the proof of Theorem 4. Then for every $m \geq 1$

$$\int f_m \, \mu(dx) = \lim_{n\to\infty} \int f_m \, \mu_n(dx) \geq \overline{\lim_{n\to\infty}} \, \mu_n(\Gamma)$$

since $f_m \geq I_\Gamma$. In addition, the left hand sides converge to $\mu(\Gamma)$ as $m \to \infty$, so that $\mu(\Gamma) \geq \overline{\lim_{n\to\infty}} \, \mu_n(\Gamma)$. The second equality in (ii) is obvious since $\int 1 \, \mu_n(dx) \to \int 1 \, \mu(dx)$.

Obviously (ii)$\Longleftrightarrow$(iii).

(ii)&(iii)$\Longrightarrow$(iv). Indeed,

$$\bar{B} \supset B \supset \bar{B} \setminus \partial B,$$

where $\bar{B}$ is closed, $\bar{B} \setminus \partial B$ is open, $\partial B \subset \bar{B}$, $\mu(\bar{B} \setminus (\bar{B} \setminus \partial B)) = \mu(\partial B) = 0$. Hence

$$\mu(\bar{B}) = \mu(\bar{B} \setminus \partial B) = \mu(B)$$

and

$$\mu(B) = \mu(\bar{B}) \geq \overline{\lim_{n\to\infty}} \, \mu_n(\bar{B}) \geq \overline{\lim_{n\to\infty}} \, \mu_n(B) \geq \underline{\lim_{n\to\infty}} \, \mu_n(B)$$

$$\geq \underline{\lim_{n\to\infty}} \, \mu_n(\bar{B} \setminus \partial B) \geq \mu(\bar{B} \setminus \partial B) = \mu(B).$$

(iv)$\Longrightarrow$(v). First, since $\partial X = \emptyset$, $\mu_n(X) \to \mu(X)$. It follows that we can add any constant to f without altering (4), which allows us to concentrate only on $f \geq 0$. For such a bounded f we have

$$\int f \, \mu_n(dx) = \int \Big(\int_0^M I_{f(x)>t} \, dt \Big) \, \mu_n(dx) = \int_0^M \mu_n\{x : f(x) > t\} \, dt,$$

where $M = \sup f$. It is seen now that, to prove (4), it suffices to show that

$$\mu_n\{x : f(x) > t\} \to \mu\{x : f(x) > t\} \tag{5}$$

for almost all t. We will see that this convergence holds at every point t at which $\mu\{x : f(x) = t\} = 0$; that is, one needs to exclude not more than a countable set.

Take a $t > 0$ such that $\mu\{x : f(x) = t\} = 0$ and let $B = \{x : f(x) > t\}$. If $y \in \partial B$ and f is continuous at y, then $f(y) = t$. Hence $\partial B \subset \{f(x) = t\} \cup \Delta_f$, $\mu(\partial B) = 0$, and (5) follows from the assumption.

Finally, since the implication (v)$\Longrightarrow$(i) is obvious, the theorem is proved.

Before stating the following corollary we remind the reader that we have defined weak convergence (Definition 8) only for nonnegative finite measures.

12. Corollary. *Let X be a closed subset in $\mathbb{R}^d$ and $\mu_n \xrightarrow{w} \ell$, where ℓ is Lebesgue measure. Then (4) holds for every Borel Riemann integrable function f, since for such a function $\ell(\Delta_f) = 0$.*

13. Exercise. If α is an irrational number in $(0,1)$, then, for every integer $m \neq 0$ and every $x \in \mathbb{R}$,

$$\frac{1}{n+1} \sum_{k=0}^{n} e^{im2\pi(x+k\alpha)} = e^{im2\pi x} \frac{e^{im2\pi(n+1)\alpha} - 1}{(n+1)(e^{im2\pi\alpha} - 1)} \to 0 \quad \text{as} \quad n \to \infty. \tag{6}$$

Also, if $m = 0$, the limit is just 1. By using Fourier series, prove that

$$\frac{1}{n+1} \sum_{k=0}^{n} f(x + k\alpha) \to \int_0^1 f(y)\, dy \tag{7}$$

for every $x \in [0,1]$ and every 1-periodic continuous function f. By writing the sum in (7) as the integral against a measure μ_n and applying Corollary 12 for indicators, prove that, for every $0 \leq a < b \leq 1$, the asymptotic frequency of fractional parts of numbers $\alpha, 2\alpha, 3\alpha, ...$ in the interval (a, b) is $b - a$.

14. Exercise. Take the sequence 2^n, $n = 1, 2, ...$, and, for each n, let a_n be the first digit in the decimal form of 2^n. Here is the sequence of the first 45 values of a_n obtained by using Matlab:

$$2, 4, 8, 1, 3, 6, 1, 2, 5, 1, 2, 4, 8, 1, 3, 6, 1, 2, 5, 1, 2, 4,$$

$$8, 1, 3, 6, 1, 2, 5, 1, 2, 4, 8, 1, 3, 6, 1, 2, 5, 1, 2, 4, 8, 1, 3.$$

We see that there are no 7s or 9s in this sequence. Let $N_b(n)$ denote the number of appearances of digit $b = 1, ..., 9$ in the sequence $a_1, ..., a_n$. By using Exercise 13 find the limit of $N_b(n)/n$ as $n \to \infty$ and, in particular, show that this limit is positive for every $b = 1, ..., 9$.

15. Exercise. Prove that for every function f (measurable or not) the set Δ_f is Borel.

We will use the following theorem, the proof of which can be found in [**Bi**].

16. Theorem (Prokhorov). *A family $\mathcal{M}$ of probability measures on the space $(X, \mathfrak{B}(X))$ is relatively weakly compact if and only if it is tight in the sense that for every $\varepsilon > 0$ there exists a compact set K such that $\mu(K^c) \le \varepsilon$ for* every *$\mu \in \mathcal{M}$.*

Let us give an outline of a proof of this theorem (a complete proof can be found, for instance, in [**Bi**], [**Du**], [**GS**]). The necessity is proved in the same way as Ulam's theorem. Indeed, use the notation from its proof and first prove that for every $n \ge 1$

$$\inf\{\mu(\bigcup_{i \le m} B_{1/n}(x_i)) : \mu \in \mathcal{M}\} \to 1 \tag{8}$$

as $m \to \infty$. By way of getting a contradiction, assume that this is wrong. Then, for an $\varepsilon > 0$ and every m, there would exist a measure $\mu_m \in \mathcal{M}$ such that

$$\mu_m(\bigcup_{i \le m} B_{1/n}(x_i)) \le 1 - \varepsilon.$$

By assumption there exists a (probability) measure μ which is a weak limit point of $\{\mu_m\}$. By Ulam's theorem there is a compact set K such that $1 - \varepsilon/2 \le \mu(K)$. Since K admits a finite $1/(2n)$-net, there exists k such that $K \subset \bigcup_{i \le k} B^o_{1/n}(x_i)$, where $B^o_r(x)$ is the open ball of radius r centered at x. By Theorem 11 (iii)

$$1 - \varepsilon/2 \le \mu(\bigcup_{i \le k} B^o_{1/n}(x_i)) \le \varliminf_{m \to \infty} \mu_m(\bigcup_{i \le k} B^o_{1/n}(x_i))$$

$$\le \varlimsup_{m \to \infty} \mu_m(\bigcup_{i \le k} B_{1/n}(x_i)) \le 1 - \varepsilon.$$

We have a contradiction which proves (8). Now it is clear how to choose i_n in order to have (2) satisfied for all $\mu \in \mathcal{M}$, and then the desired set K can be given by (3).

Proof of sufficiency can be based on Riesz's remarkable theorem on the general form of continuous linear functionals defined on the set of continuous functions on a compact set. Let K be a compact subset of X, $C(K)$ the set of all continuous functions on K, and assume that on $C(K)$ we have a linear function $\ell(f)$ such that

$$|\ell(f)| \le N \sup\{|f(x)| : x \in K\}$$

for all $f \in C(K)$ with N independent of f. Then it turns out that there is a measure μ such that

$$\ell(f) = \int_K f\, \mu(dx).$$

Now fix $\varepsilon > 0$ and take an appropriate $K(\varepsilon)$. For every countable set of $f_m \in C(K(\varepsilon))$ and every sequence of measures $\mu \in \mathcal{M}$, by using Cantor's diagonalization method one can extract a subsequence μ_n such that

$$\int_{K(\varepsilon)} f_m\, \mu_n(dx)$$

would have limits as $n \to \infty$. One can choose such a sequence of f's to be dense in $C(K(\varepsilon))$, and then $\lim_{n\to\infty} \int_{K(\varepsilon)} f\, \mu_n(dx)$ exists for every continuous f and defines a linear bounded functional on $C(K(\varepsilon))$, and hence defines a measure on $K(\varepsilon)$. It remains to paste these measures obtained for different ε and get a measure on X, and also arrange for one sequence μ_n to be good for all $K(\varepsilon)$ with ε running through $1, 1/2, 1/3, ...$

17. Remark. In the above explanation we used the fact that if μ is a finite measure on $(X, \mathfrak{B}(X))$ and $\int f\, \mu(dx) \geq 0$ for all nonnegative bounded continuous functions f, then $\mu \geq 0$.

To prove that this is indeed true, remember that by Hahn's theorem there exist two measurable (Borel) sets B_1 and B_2 such that $B_1 \cup B_2 = X$, $B_1 \cap B_2 = \emptyset$, and $\mu_i(B) = (-1)^i \mu(B \cap B_i) \geq 0$ for $i = 1, 2$ and every $B \in \mathfrak{B}(X)$.

Then $\mu = \mu_2 - \mu_1$, $\int f\, \mu_2(dx) \geq \int f\, \mu_1(dx)$ for all nonnegative continuous f. One derives from here as in the proof of Theorem 4 that $\mu_2(\Gamma) \geq \mu_1(\Gamma)$ for all closed Γ, and by regularity $\mu_2(B) \geq \mu_1(B)$ for all $B \in \mathfrak{B}(X)$. Plugging in $B \cap B_1$ in place of B, we get $0 = \mu_2(B \cap B_1) \geq \mu_1(B \cap B_1) = \mu_1(B) \geq 0$ and $\mu_1(B) = 0$, as claimed.

3. The notion of random process

Let T be a set, $(\Omega, \mathcal{F}, P)$ a probability space, $(X, \mathfrak{B})$ a measurable space, and assume that, for every $t \in T$, we are given an X-valued $\mathcal{F}$-measurable function $\xi_t = \xi_t(\omega)$. Then we say that ξ_t is *a random process* on T with values in X. For individual ω the function $\xi_t(\omega)$ as a function of t is called *a path* or *a trajectory of the process*.

The set T may be different in different settings. If $T = \{0, 1, 2, ...\}$, then ξ_t is called *a random sequence*. If $T = (a, b)$, then ξ_t is a *continuous-time random process*. If $T = \mathbb{R}^2$, then ξ_t is called a two-parameter *random field*.

In the following lemma, for a measurable space $(X, \mathfrak{B})$ and integer n, we denote by $(X^n, \mathfrak{B}^n)$ the product of n copies of $(X, \mathfrak{B})$, that is,

$$X^n = \{(x^1, ..., x^n) : x^1, ..., x^n \in X\},$$

and $\mathfrak{B}^n$ is the smallest σ-field of subsets of X^n containing every $B^{(n)}$ of type

$$B_1 \times ... \times B_n,$$

where $B_i \in \mathfrak{B}(X)$.

1. Lemma. *Let $t_1, ..., t_n \in T$. Then $(\xi_{t_1}, ..., \xi_{t_n})$ is a random variable with values in $(X^n, \mathfrak{B}^n)$.*

Proof. The function $\eta(\omega) := (\xi_{t_1}(\omega), ..., \xi_{t_n}(\omega))$ maps Ω into X^n. The set Σ of all subsets $B^{(n)}$ of X^n for which $\eta^{-1}(B^{(n)}) \in \mathcal{F}$ is a σ-field. In addition, Σ contains every $B^{(n)}$ of type $B_1 \times ... \times B_n$, where $B_i \in \mathfrak{B}(X)$. This is seen from the fact that

$$\eta^{-1}(B_1 \times ... \times B_n) = \{\omega : \eta(\omega) \in B_1 \times ... \times B_n\}$$

$$= \{\omega : \xi_{t_1}(\omega) \in B_1, ..., \xi_{t_n}(\omega) \in B_n\} = \bigcap_i \{\omega : \xi_{t_i}(\omega) \in B_i\} \in \mathcal{F}.$$

Hence Σ contains the σ-field generated by those $B^{(n)}$. Since the latter is $\mathfrak{B}^n$ by definition, we have $\Sigma \supset \mathfrak{B}(X)$, i.e. $\eta^{-1}(B^{(n)}) \in \mathcal{F}$ for every $B^{(n)} \in \mathfrak{B}^n$. The lemma is proved.

2. Remark. In particular, we have proved that $\{\omega : \xi^2(\omega) + \eta^2(\omega) \leq 1\}$ is a random event if ξ and η are random variables.

The random variable $(\xi_{t_1}, ..., \xi_{t_n})$ has a distribution on $(X^n, \mathfrak{B}^n)$. This distribution is called *the finite-dimensional distribution* corresponding to $t_1, ..., t_n$.

So-called cylinder sets play an important role in the theory of random processes.

Let $(X, \mathfrak{B}(X))$ be a Polish space and T a set. Denote by X^T the set of all X-valued functions on T. This notation is natural if one observes that if T only consists of two points, $T = \{1, 2\}$, then every X-valued function on T is just a pair (x, y), where x is the value of the function at $t = 1$ and y is the value of the function at $t = 2$. So the set of X-valued functions on T is just the set of all pairs (x, y), and $X^T = X \times X = X^2$.

We denote by $x.$ the points in X^T and by x_t the value of $x.$ at t. Every set of type

$$\{x. : (x_{t_1}, ..., x_{t_n}) \in B^{(n)}\},$$

where $t_i \in T$ and $B^{(n)} \in \mathfrak{B}^n$, is called *the finite dimensional cylinder set* with base $B^{(n)}$ attached to $t_1, ..., t_n$. The σ-field generated by all finite dimensional cylinder sets is called *the cylinder σ-field*.

3. Exercise*. Prove that the family of all finite dimensional cylinder sets is an algebra, that is, X^T is a cylinder set and complements and finite unions and intersections of cylinder sets are cylinder sets.

4. Exercise. Let Σ denote the cylinder σ-field of subsets of the set of all X-valued functions on $[0, 1]$. Prove that for every $A \in \Sigma$ there exists a countable set $t_1, t_2, ... \in [0, 1]$ such that if $x. \in A$ and $y.$ is a function such that $y_{t_n} = x_{t_n}$ for all n, then $y. \in A$. In other words, elements of Σ are defined by specifying conditions on trajectories only at countably many points of $[0, 1]$.

5. Exercise. Give an example of a Polish space $(X, \mathfrak{B}(X))$ such that the set $C([0, 1], X)$ of all bounded and continuous X-valued functions on $[0, 1]$ is not an element of the σ-field Σ from the previous exercise. Thus you will see that there exists a very important and quite natural set which is not measurable.

4. Continuous random processes

For simplicity consider real-valued random processes on $T = [0, 1]$. Such a process is called *continuous* if all its trajectories are continuous functions on T. In that case, for each ω, we have a continuous trajectory or in other words an element of the space $C = C([0, 1])$ of continuous functions on $[0, 1]$. You know that this is a Polish space when provided with the metric

$$\rho(x., y.) = \sup_{t \in [0,1]} |x_t - y_t|.$$

Apart from the Borel σ-field, which is convenient as far as convergence of distributions is concerned, there is *the cylinder σ-field* $\Sigma(C)$, defined as the σ-field of subsets of C generated by the collection of all subsets of the form

$$\{x. \in C : x_t \in \Gamma\}, \quad t \in [0, 1], \Gamma \in \mathfrak{B}(\mathbb{R}).$$

Observe that $\Sigma(C)$ is *not* the cylinder σ-field in the space of all real-valued functions on $[0, 1]$ as defined before Exercise 3.3.

1. Lemma. $\Sigma(C) = \mathfrak{B}(C)$.

Proof. For t fixed, denote by π_t the function on C defined by

$$\pi_t(x.) = x_t.$$

Obviously π_t is a real-valued continuous function on C. By Theorem 2.1 it is Borel, i.e. for every $B \in \mathfrak{B}(\mathbb{R})$ we have $\pi_t^{-1}(B) \in \mathfrak{B}(C)$, i.e. $\{x_\cdot : x_t \in B\} \in \mathfrak{B}(C)$. It follows easily (for instance, as in the proof of Theorem 2.1) that $\Sigma(C) \subset \mathfrak{B}(C)$.

To prove the opposite inclusion it suffices to prove that all closed balls are cylinder sets. Fix $x^0_\cdot \in C$ and $\varepsilon > 0$. Then obviously

$$B_\varepsilon(x^0_\cdot) = \{x_\cdot \in C : \rho(x^0_\cdot, x_\cdot) \leq \varepsilon\} = \bigcap \{x_\cdot \in C : x_r \in [x^0_r - \varepsilon, x^0_r + \varepsilon]\},$$

where the intersection is taken for all rational $r \in [0,1]$. This intersection being countable, we have $B_\varepsilon(x^0_\cdot) \in \Sigma(C)$, and the lemma is proved.

The following theorem allows one to treat continuous random processes just like C-valued random elements.

2. Theorem. *If $\xi_t(\omega)$ is a continuous process on $[0,1]$, then $\xi_\cdot$ is a C-valued random variable. Conversely, if $\xi_\cdot$ is a C-valued random variable, then $\xi_t(\omega)$ is a continuous process on $[0,1]$.*

Proof. To prove the direct statement, it suffices to notice that, by definition, the σ-field of all those $B \subset C$ for which $\xi_\cdot^{-1}(B) \in \mathcal{F}$ contains all sets of the type

$$\{x_\cdot : x_t \in \Gamma\}, \quad t \in [0,1], \Gamma \in \mathfrak{B}(\mathbb{R}),$$

and hence contains all cylinder subsets of C, that is, by Lemma 1, all Borel subsets of C.

The converse follows at once from the fact that $\xi_t = \pi_t(\xi_\cdot)$, which shows that ξ_t is a superposition of two measurable functions. The lemma is proved.

By Ulam's theorem the distribution of a process with continuous trajectories is concentrated up to ε on a compact set $K_\varepsilon \subset C$. Remember the following necessary and sufficient condition for a subset of C to be compact (the Arzelà-Ascoli theorem).

3. Theorem. *Let K be a closed subset of C. It is compact if and only if the family of functions $x_\cdot \in K$ is uniformly bounded and equicontinuous, i.e. if and only if*

(i) *there is a constant N such that*

$$\sup_t |x_t| \leq N \quad \forall x_\cdot \in K$$

and

(ii) *for each $\varepsilon > 0$ there exists a $\delta > 0$ such that $|x_t - x_s| \leq \varepsilon$ whenever $x_\cdot \in K$ and $|t - s| \leq \delta$, $t, s \in [0,1]$.*

4. Lemma. *Let x_t be a real-valued function defined on $[0,1]$ (independent of ω). Assume that there exist a constant $a > 0$ and an integer $n \geq 0$ such that*

$$|x_{(i+1)/2^m} - x_{i/2^m}| \leq 2^{-ma}$$

for all $m \geq n$ and $0 \leq i \leq 2^m - 1$. Then for all binary rational numbers $t, s \in [0,1]$ satisfying $|t - s| \leq 2^{-n}$ we have

$$|x_t - x_s| \leq N(a)|t - s|^a,$$

where $N(a) = 2^{2a+1}(2^a - 1)^{-1}$.

Proof. Let $t, s \in [0,1]$ be binary rational. Then

$$t = \sum_{i=0}^{\infty} \varepsilon_1(i) 2^{-i}, \quad s = \sum_{i=0}^{\infty} \varepsilon_2(i) 2^{-i}, \tag{1}$$

where $\varepsilon_k(i) = 0$ or 1 and the series are actually finite sums. Let

$$t_k = \sum_{i=0}^{k} \varepsilon_1(i) 2^{-i}, \quad s_k = \sum_{i=0}^{k} \varepsilon_2(i) 2^{-i}. \tag{2}$$

Observe that if $|t - s| \leq 2^{-k}$, then $t_k = s_k$ or $|t_k - s_k| = 2^{-k}$. This follows easily from the following picture in which $|$ shows numbers of type $r2^{-k}$ with integral r, the short arrow shows the set of possible values for t and the long one the set of possible values of s.

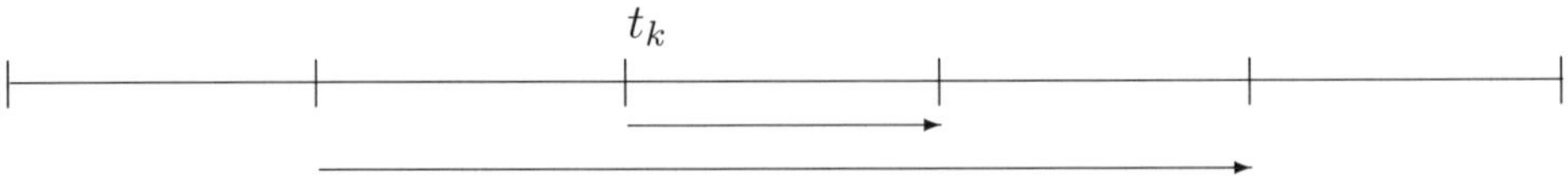

Now let $k \geq n$ and $|t - s| \leq 2^{-k}$. Write

$$x_t = x_{t_k} + \sum_{m=k}^{\infty} (x_{t_{m+1}} - x_{t_m}),$$

write similar representation for x_s and subtract these formulas to get

$$|x_t - x_s| \leq |x_{t_k} - x_{s_k}| + \sum_{m=k}^{\infty} \{|x_{t_{m+1}} - x_{t_m}| + |x_{s_{m+1}} - x_{s_m}|\}. \tag{3}$$

Here $t_k = r2^{-k}$ for an integer r, and there are only three possibility for s_k: $s_k = (r-1)2^{-k}$ or $= r2^{-k}$ or $= (r+1)2^{-k}$. In addition, $|t_{m+1}-t_m| \leq 2^{-(m+1)}$ since, for an integer p, we have $t_m = p2^{-m} = (2p)2^{-(m+1)}$ and t_{m+1} equals either t_m or $t_m + 2^{-(m+1)}$. Therefore, by the assumption,

$$|x_t - x_s| \leq 2\sum_{m=k}^{\infty} 2^{-ma} = 2^{-ka}2^{a+1}(2^a-1)^{-1}. \tag{4}$$

We have proved this inequality if

$$k \geq n \quad \text{and} \quad |t-s| \leq 2^{-k}.$$

It is easy to prove that, for every t and s satisfying $|t-s| \leq 2^{-n}$, one can take $k = [\log_2(1/|t-s|)]$ and then one has $k \geq n$, $|t-s| \leq 2^{-k}$, and $2^{-ka} \leq 2^a|t-s|^a$. This proves the lemma.

For integers $n \geq 0$ and $a > 0$ denote

$$K_n(a) = \{x_\cdot \in C : |x_0| \leq 2^n, |x_t - x_s| \leq N(a)|t-s|^a \quad \forall |t-s| \leq 2^{-n}\}.$$

5. Exercise*. Prove that $K_n(a)$ are compact sets in C.

6. Theorem. *Let ξ_t be a continuous process and let $\alpha > 0, \beta > 0, N \in (0,\infty)$ be constants such that*

$$E|\xi_t - \xi_s|^\alpha \leq N|t-s|^{1+\beta} \quad \forall s,t \in [0,1].$$

Then for $0 < a < \beta\alpha^{-1}$ and for every $\varepsilon > 0$ there exists n such that

$$P\{\xi_\cdot \in K_n(a)\} \geq 1 - \varepsilon.$$

(observe that $P\{\xi_\cdot \in K_n(a)\}$ makes sense by Theorem 2).

Proof. Denote

$$A_n = \{\omega : |\xi_0| \geq 2^n\} \cup \{\omega : \sup_{m \geq n} \max_{i=0,\dots,2^m-1} |\xi_{(i+1)/2^m} - \xi_{i/2^m}|2^{ma} > 1\}.$$

For $\omega \notin A_n$, we have $\xi_\cdot \in K_n(a)$ by the previous lemma. Hence by Chebyshev's inequality

$$P\{\xi_\cdot \notin K_n(a)\} \leq P(A_n) \leq P\{|\xi_0| \geq 2^n\}$$

$$+E \sup_{m\geq n} \max_{i=0,...,2^m-1} |\xi_{(i+1)/2^m} - \xi_{i/2^m}|^\alpha 2^{ma\alpha}.$$

We replace the sup and the max with sums of the random variables involved and we find

$$P\{\xi_\cdot \notin K_n(a)\} \leq P(A_n) \leq P\{|\xi_0| \geq 2^n\} \tag{5}$$

$$+\sum_{m=n}^{\infty} \sum_{i=0}^{2^m-1} 2^{ma\alpha} E|\xi_{(i+1)/2^m} - \xi_{i/2^m}|^\alpha \leq P\{|\xi_0| \geq 2^n\} + N \sum_{m=n}^{\infty} 2^{-m(\beta - a\alpha)}.$$

It only remains to notice that the last expression tends to zero as $n \to \infty$. The theorem is proved.

Remember that if $\xi_\cdot$ is a C-valued random variable, then the measure $P\{\xi_\cdot \in B\}$, $B \in \mathfrak{B}(C)$, is called the distribution of $\xi_\cdot$. From (5) and Prokhorov's theorem we immediately get the following.

7. Theorem. *Let ξ_t^k, $k = 1, 2, 3, ...$, be continuous processes on $[0, 1]$ such that, for some constants $\alpha > 0, \beta > 0, N \in (0, \infty)$, we have*

$$E|\xi_t^k - \xi_s^k|^\alpha \leq N|t-s|^{1+\beta} \quad \forall s, t \in [0,1], k \geq 1.$$

Also assume that $\sup_k P\{|\xi_0^k| \geq c\} \to 0$ as $c \to \infty$. Then the sequence of distributions of $\xi_\cdot^k$ on C is relatively compact.

Lemma 4 is the main tool in proving Theorems 6 and 7. It also allows us to prove Kolmogorov's theorem on existence of continuous modifications.

If T is a set on which we are given two processes ξ_t^1 and ξ_t^2 such that $P(\xi_t^1 = \xi_t^2) = 1$ for every $t \in T$, then we call $\xi_\cdot^1$ *a modification of* $\xi_\cdot^2$ (and vice versa).

8. Theorem (Kolmogorov). *Let ξ_t be a process defined for $t \in [0, \infty)$ such that, for some $\alpha > 0, \beta > 0, N < \infty$, we have*

$$E|\xi_t - \xi_s|^\alpha \leq N|t-s|^{1+\beta} \quad \forall t, s \geq 0.$$

Then the process ξ_t has a continuous modification.

Proof. Take $a = \beta/(2\alpha)$ and define

$$\Omega_{kn} = \{\omega : \sup_{m\geq n} \max_{i=0,...,k2^m-1} 2^{ma}|\xi_{(i+1)/2^m} - \xi_{i/2^m}| \leq 1\}, \quad \Omega' = \bigcap_{k\geq 1} \bigcup_n \Omega_{kn}.$$

If $\omega \in \Omega'$, then for every $k \geq 1$ there exists n such that for all $m \geq n$ and $i = 0, ..., k2^m - 1$ we have

$$|\xi_{(i+1)/2^m}(\omega) - \xi_{i/2^m}(\omega)| \leq 2^{-ma}.$$

It follows by Lemma 4 that, for $\omega \in \Omega'$ and every k, the function $\xi_t(\omega)$ is uniformly continuous on the set $\{r/2^m\}$ of binary fractions intersected with $[0, k]$. By using Cauchy's criterion, it is easy to prove that, for $\omega \in \Omega'$ and every $t \in [0, \infty)$, there exists

$$\lim_{r/2^m \to t} \xi_{r/2^m}(\omega) =: \tilde{\xi}_t(\omega),$$

and in addition, $\tilde{\xi}_t(\omega)$ is continuous and $\tilde{\xi}_t(\omega) = \xi_t(\omega)$ for all binary rational t. We have defined $\tilde{\xi}_t(\omega)$ for $\omega \in \Omega'$. For $\omega \notin \Omega'$ define $\tilde{\xi}_t(\omega) \equiv 0$. The process $\tilde{\xi}_t$ is continuous, and it only remains to prove that it is a modification of ξ_t.

First we claim that $P(\Omega') = 1$. To prove this it suffices to prove that $P(\bigcup_n \Omega_{kn}) = 1$. Since $(\bigcup_n \Omega_{kn})^c = \bigcap_n \Omega_{kn}^c$ and

$$\Omega_{kn}^c = \bigcup_{m \geq n} \bigcup_{i=0}^{k2^m - 1} \{\omega : |\xi_{(i+1)/2^m} - \xi_{i/2^m}| > 2^{-ma}\},$$

we have (cf. (5))

$$1 - P(\bigcup_n \Omega_{kn}) \leq \overline{\lim_{n \to \infty}} \sum_{m \geq n} kN 2^{m(a\alpha - \beta)} = 0.$$

Thus $P(\Omega') = 1$. Furthermore, we noticed above that $\tilde{\xi}_{r/2^m} = \xi_{r/2^m}$ on Ω'. Therefore,

$$P\{\tilde{\xi}_{r/2^m} - \xi_{r/2^m}\} = 1.$$

For other values of t, by Fatou's theorem

$$E|\tilde{\xi}_t - \xi_t|^\alpha \leq \lim_{r/2^k \to t} E|\xi_{r/2^k} - \xi_t|^\alpha \leq N \lim_{r/2^k \to t} |r/2^k - t|^{1+\beta} = 0.$$

Hence $P\{\tilde{\xi}_t = \xi_t\} = 1$ for every $t \in [0, \infty)$, and the theorem is proved.

For Gaussian processes the above results can be improved. Remember that a random vector $\xi = (\xi_1, ..., \xi_k)$ with values in $\mathbb{R}^k$ is called *Gaussian* or *normal* if there exist a vector $m \in \mathbb{R}^k$ and a symmetric nonnegative $k \times k$ matrix $R = (R_{ij})$ such that

$$\varphi(\lambda) := E \exp(i(\xi, \lambda)) = \exp(i(\lambda, m) - (R\lambda, \lambda)/2) \quad \forall \lambda \in \mathbb{R}^k,$$

where

$$(\lambda, \mu) = \sum_{i=1}^{k} \lambda_i \mu_i$$

is the scalar product in $\mathbb{R}^k$ and

$$(R\lambda, \lambda) = \sum_{i,j=1}^{k} R_{ij} \lambda_i \lambda_j .$$

In this case one also writes $\xi \sim N(m, R)$. One knows that

$$m = E\xi, \quad R_{ij} = E(\xi_i - m_i)(\xi_j - m_j),$$

so that m is the mean value of ξ and R is its covariance matrix. It is known that linear transformations of Gaussian vectors are Gaussian. In particular, $(\xi_2, \xi_1, \xi_3, ..., \xi_k)$ is Gaussian.

9. Definition. A real-valued process ξ_t is called *Gaussian* if all its finite-dimensional distributions are Gaussian. The function $m_t = E\xi_t$ is called *the mean value function* of ξ_t and $R(t, s) = \operatorname{cov}(\xi_t, \xi_s) = E(\xi_t - m_t)(\xi_s - m_s)$ is called *the covariance function* of ξ_t.

10. Remark. Very often it is useful to remember that $(x_{t_1}, ..., x_{t_k})$ is a k-dimensional Gaussian vector if an only if, for arbitrary constants $c_1, ..., c_k$, the random variable $\sum_i c_i x_{t_i}$ is Gaussian.

11. Exercise. Let x_t be a real-valued function defined on $[0, 1]$ (independent of ω). Let $g(x)$ be a nonnegative increasing function defined on $(0, 1/2]$ and such that

$$G(x) = \int_0^x y^{-1} g(y) \, dy$$

is finite for every $x \in [0, 1/2]$. Assume that there exists an integer $n \geq 3$ such that

$$|x_{(i+1)/2^m} - x_{i/2^m}| \leq g(2^{-m})$$

for all $m \geq n$ and $0 \leq i \leq 2^m - 1$. By analyzing the proof of Lemma 4, show that for all binary rational numbers $t, s \in [0, 1]$ satisfying $|t - s| \leq 2^{-n}$ we have

$$|x_t - x_s| \leq NG(4|t - s|), \quad N = 2/\ln 2 .$$

12. Exercise. Let ξ be a normal random variable with zero mean and variance less than or equal to σ^2, where $\sigma > 0$. Prove that, for every $x > 0$,

$$\sqrt{2\pi} \, P(|\xi| \geq x) \leq 2\sigma x^{-1} \exp(-x^2/(2\sigma^2)).$$

13. Exercise. Let ξ_t be a Gaussian process with zero mean given on $[0,1]$ and satisfying $E|\xi_t - \xi_s|^2 \leq R(|t-s|)$, where R is a continuous function defined on $(0,1]$. Denote $g(x) = \sqrt{R(x)(-\ln x)}$ and suppose that g satisfies the assumptions of Exercise 11. For a constant $a > \sqrt{2}$ and $n \geq 3$ define

$$\Omega_n = \{\omega : \sup_{m\geq n} \max_{i=0,\dots,2^m-1} |\xi_{(i+1)/2^m}(\omega) - \xi_{i/2^m}(\omega)|/g(2^{-m}) \leq a\},$$

$\Omega' = \bigcup_{n\geq 3} \Omega_n$. Notice that

$$\Omega_n^c = \bigcup_{m=n}^{\infty} \bigcup_{i=0}^{2^m-1} \{\omega : |\xi_{(i+1)/2^m}(\omega) - \xi_{i/2^m}(\omega)| > ag(2^{-m})\}$$

and, by using Exercise 12, prove that

$$P(\Omega_n^c) \leq N_1 \sum_{m\geq n} 2^m \frac{\sqrt{R(2^{-m})}}{g(2^{-m})} \exp\Big(-\frac{a^2 g^2(2^{-m})}{2R(2^{-m})}\Big)$$
$$= N_2 \sum_{m\geq n} \frac{1}{\sqrt{m}} 2^{m(1-a^2/2)},$$

where the N_i are independent of n. Conclude that $P(\Omega') = 1$. By using Exercise 11, derive from here that ξ_t has a continuous modification. In particular, prove that, if

$$E|\xi_t - \xi_s|^2 \leq N(-\ln|t-s|)^{-p} \quad \forall t,s \in [0,1], |t-s| \leq 1/2$$

with a constant N and $p > 3$, then ξ_t has a continuous modification.

14. Exercise. Let ξ_t be a process satisfying the assumptions in Exercise 13 and let $\tilde{\xi}_t$ be its continuous modification. Prove that, for almost every ω, there exists $n \geq 1$ such that for all $t,s \in [0,1]$ satisfying $|t-s| \leq 2^{-n}$ we have

$$|\tilde{\xi}_t - \tilde{\xi}_s| \leq 8G(4|t-s|). \qquad (G(x) = \int_0^x y^{-1} g(y)\,dy)$$

Sometimes one needs the following multidimensional version of Kolmogorov's theorem. To prove it we first generalize Lemma 4. Denote by $\mathbb{Z}_n^d$ the lattice in $[0,1]^d$ consisting of all points $(k_1 2^{-n}, \dots, k_d 2^{-n})$ where $k_i = 0,1,2,\dots,2^n$. Also let

$$||t-s|| = \max\{|t^i - s^i| : i = 1,\dots,d\}.$$

15. Lemma. *Let $d \geq 1$ be an integer and x_t a real-valued function defined for $t \in [0,1]^d$. Assume that there exist $a > 0$ and an integer $n \geq 0$ such that*

$$m \geq n, \quad t, s \in \mathbb{Z}^d_m, \quad ||t - s|| \leq 2^{-m} \Longrightarrow |x_t - x_s| \leq 2^{-ma}.$$

Then, for every $t, s \in \bigcup_m \mathbb{Z}^d_m$ satisfying $||t - s|| \leq 2^{-n}$ we have

$$|x_t - x_s| \leq N(a)||t - s||^a.$$

Proof. Let $t, s \in \mathbb{Z}^d_m$, $t = (t^1, ..., t^d)$, $s = (s^1, ..., s^d)$. Represent t^j and s^j as (cf. (1))

$$t^j = \sum_{i=0}^{\infty} \varepsilon^j_1 2^{-i}, \quad s^j = \sum_{i=0}^{\infty} \varepsilon^j_2 2^{-i},$$

define t^j_k and s^j_k as these sums for $i \leq k$ (cf. (2)), and let

$$t_k = (t^1_k, ..., t^d_k), \quad s_k = (s^1_k, ..., s^d_k).$$

Then $||t - s|| \leq 2^{-k}$ implies $|t^j_k - s^j_k| \leq 2^{-k}$, and as in Lemma 4 we get $t_k, s_k \in \mathbb{Z}^d_k$, $|t^j_k - s^j_k| \leq 2^{-k}$ and $||t_k - s_k|| \leq 2^{-k}$. We use (3) again and the fact that, as before, $t_{m+1}, t_m \in \mathbb{Z}^d_{m+1}$, $||t_{m+1} - t_m|| \leq 2^{-(m+1)}$. Then we get (4) again and finish the proof by the same argument as before. The lemma is proved.

Now we prove a version of Theorem 6. For an integer $n \geq 0$ denote

$$\Gamma_n(a) = \{x_\cdot : x_t \text{ is a real-valued function given on } [0,1]^d \text{ such that}$$
$$|x_t - x_s| \leq N(a)||t - s||^a \quad \text{for all } t, s \in \bigcup_m \mathbb{Z}^d_m \text{ with } ||t - s|| \leq 2^{-n}\}.$$

16. Lemma. *Let a random field ξ_t be defined on $[0,1]^d$. Assume that there exist constants $\alpha > 0, \beta > 0$, $K < \infty$ such that*

$$E|\xi_t - \xi_s|^\alpha \leq K||t - s||^{d+\beta}$$

provided $t, s \in [0,1]^d$. Then, for every $0 < a < \beta/\alpha$,

$$P\{\xi_\cdot \notin \Gamma_n(a)\} \leq 2^{-n(\beta - a\alpha)} K N(d, \alpha, \beta, a).$$

Proof. Let

$$A_n = \{\omega : \sup_{m \geq n} \sup\{2^{ma}|\xi_t - \xi_s| : t, s \in \mathbb{Z}^d_m, ||t - s|| \leq 2^{-m}\} > 1\}.$$

For $\omega \notin A_n$ we get $\xi_\cdot(\omega) \in \Gamma_n(a)$ by Lemma 15. Hence, $P\{\xi_\cdot \notin \Gamma_n(a)\} \leq P(A_n)$. The probability of A_n we again estimate by Chebyshev's inequality and estimate the α power of the sup through the sum of α powers of the random variables involved. For each m the number of these random variables is not greater than the number of couples $t, s \in \mathbb{Z}^d_m$ for which $||t - s|| \leq 2^{-m}$ (and the number of disjoint ones is less than half this number). This number is not bigger than the number of points in $\mathbb{Z}^d_m$ times 3^d, the latter being the number of neighbors of t. Hence

$$P(A_n) \leq \sum_{m=n}^{\infty} (1 + 2^m)^d 3^d K 2^{ma\alpha} 2^{-m(d+\beta)} \leq 6^d K \sum_{m=n}^{\infty} 2^{-m(\beta - a\alpha)}$$

$$= 2^{-n(\beta - a\alpha)} K 6^d (1 - 2^{-(\beta - a\alpha)})^{-1}.$$

The lemma is proved.

17. Theorem (Kolmogorov). *Under the conditions of Lemma* 16 *the random field* ξ_t *has a continuous modification.*

Proof. By Lemma 16, with probability one, $\xi_\cdot$ belongs to one of the sets $\Gamma_n(a)$. The elements of these sets are uniformly continuous on $\bigcup_m \mathbb{Z}^d_m$ and therefore can be redefined outside $\bigcup_m \mathbb{Z}^d_m$ to become continuous on $[0, 1]^d$. Hence, with probability one there exists a continuous function $\tilde{\xi}_t$ coinciding with ξ_t on $\bigcup_m \mathbb{Z}^d_m$. To finish the proof it suffices to repeat the end of the proof of Theorem 8. The theorem is proved.

5. Hints to exercises

1.7 It suffices to prove that $A^c \in \mathcal{F}^P$ if $P(A) = 0$.

2.6 To prove (i)$\Longrightarrow$(ii), observe that, for every $k \geq 1$, in the $1/k$-neighborhood of a point from a $1/k$-net there are infinitely many elements of x_n, which allows one to choose a Cauchy subsequence. To prove (ii)$\Longrightarrow$(i), assume that for an $\varepsilon > 0$ there is no finite ε-net, and find a sequence of $x_n \in K$ such that $\rho(x_n, x_m) \geq \varepsilon/3$ for all n, m.

2.10 Assume the contrary.

2.14 Observe that $N_b(n)$ is the number of $i = 1, ..., n$ such that $10^k b \leq 2^i < 10^k(b + 1)$ for some $k = 0, 1, 2, ...$, and then take $\log_{10}$.

2.15 Define

$$\bar{f}(x) = \lim_{\varepsilon\downarrow 0} \sup_{y:|y-x|<\varepsilon} f(y), \quad \underline{f}(x) = \lim_{\varepsilon\downarrow 0} \inf_{y:|y-x|<\varepsilon} f(y)$$

and prove that $\Delta_f = \{\bar{f} \neq \underline{f}\}$ and the sets $\{x : \bar{f}(x) < c\}$ and $\{x : \underline{f}(x) > c\}$ are open.

3.3 Attached points $t_1, ..., t_n$ and n may vary and $t_1, ..., t_n$ are not supposed to be distinct.

3.4 Show that the set of all such A is a σ-field.

4.12 Let $\alpha^2 = E\xi^2$. Observe that $P(|\xi| \geq x) = P(|\xi/\alpha| \geq x/\alpha)$. Then in the integral $\int_{x/\alpha}^{\infty} \exp(-y^2/2)\,dy$ first replace α with σ and after that divide and multiply the integrand by y.

Chapter 2

The Wiener Process

1. Brownian motion and the Wiener process

Robert Brown, an English botanist, observed (1828) that pollen grains suspended in water perform an unending chaotic motion. L. Bachelier (1900) derived the law governing the position w_t at time t of a single grain performing a one-dimensional Brownian motion starting at $a \in \mathbb{R}$ at time $t = 0$:

$$P_a\{w_t \in dx\} = p(t, a, x)\, dx, \tag{1}$$

where

$$p(t, a, x) = \frac{1}{\sqrt{2\pi t}} e^{-(x-a)^2/(2t)}$$

is the fundamental solution of the heat equation

$$\frac{\partial u}{\partial t} = \frac{1}{2}\frac{\partial^2 u}{\partial a^2}.$$

Bachelier (1900) also pointed out the Markovian nature of the Brownian path and used it to establish the law of maximum displacement

$$P_a\{\max_{s \le t} w_s \le b\} = \frac{2}{\sqrt{2\pi t}} \int_0^b e^{-x^2/(2t)}\, dx, \quad t > 0, b \ge 0.$$

Einstein (1905) also derived (1) from statistical mechanics considerations and applied it to the determination of molecular diameters. Bachelier was unable to obtain a clear picture of the Brownian motion, and his ideas were unappreciated at the time. This is not surprising, because the precise mathematical definition of the Brownian motion involves a measure on the path space, and even after the ideas of Borel, Lebesgue, and Daniell appeared, N. Wiener (1923) only constructed a Daniell integral on the path space which later was revealed to be the Lebesgue integral against a measure, the so-called Wiener measure.

The simplest model describing movement of a particle subject to hits by much smaller particles is the following. Let η_k, $k = 1, 2, ...$, be independent identically distributed random variables with $E\eta_k = 0$ and $E\eta_k^2 = 1$. Fix an integer n, and at times $1/n, 2/n, ...$ let our particle experience instant displacements by $\eta_1 n^{-1/2}$, $\eta_2 n^{-1/2}, ...$ At moment zero let our particle be at zero. If

$$S_k := \eta_1 + ... + \eta_k,$$

then at moment k/n our particle will be at the point $S_k/\sqrt{n}$ and will stay there during the time interval $[k/n, (k+1)/n)$. Since real Brownian motion has continuous paths, we replace our piecewise constant trajectory by a continuous piecewise linear one preserving its positions at times k/n. Thus we come to the process

$$\xi_t^n := S_{[nt]}/\sqrt{n} + (nt - [nt])\eta_{[nt]+1}/\sqrt{n}. \tag{2}$$

This process gives a very rough caricature of Brownian motion. Clearly, to get a better model we have to let $n \to \infty$. By the way, precisely this necessity dictates the intervals of time between collisions to be $1/n$ and the displacements due to collisions to be $\eta_k/\sqrt{n}$, since then ξ_t^n is asymptotically normal with parameters $(0, 1)$.

It turns out that under a very special organization of randomness, which generates different $\{\eta_k; k \geq 1\}$ for different n, one can get the situation where the ξ_t^n converge for each ω uniformly on each finite interval of time. This is a consequence of a very general result due to Skorokhod. We do not use this result, confining ourselves to the weak convergence of the distributions of $\xi_\cdot^n$.

1. Lemma. *The sequence of distributions of $\xi_\cdot^n$ in C is relatively compact.*

Proof. For simplicity we assume that $m_4 := E\eta_k^4 < \infty$, referring the reader to [**Bi**] for the proof in the general situation. Since $\xi_0^n = 0$, by Theorem 1.4.7 it suffices to prove that

$$E|\xi_t^n - \xi_s^n|^4 \leq N|t - s|^2 \quad \forall s, t \in [0, 1], \tag{3}$$

where N is independent of n, t, s.

Without loss of generality, assume that $s < t$. Denote $a_n = E(S_n)^4$. By virtue of the independence of the η_k and the conditions $E\eta_k = 0$ and $E\eta_k^2 = 1$, we have

$$a_{n+1} = E(S_n + \eta_{n+1})^4 = a_n + 4ES_n^3\eta_{n+1} + 6ES_n^2\eta_{n+1}^2 + 4ES_n\eta_{n+1}^3 + m_4 = a_n + 6n + m_4.$$

Hence (for instance, by induction),

$$a_n = 3n(n-1) + nm_4 \le 3n^2 + nm_4.$$

Furthermore, if s and t belong to the same interval $[k/n, (k+1)/n]$, then

$$|\xi_t^n - \xi_s^n| = \sqrt{n}|\eta_{k+1}|\,|t-s|,$$

$$E|\xi_t^n - \xi_s^n|^4 = n^2 m_4|t-s|^4 \le m_4|t-s|^2. \tag{4}$$

Now, consider the following picture, where s and t belong to different intervals of type $[k/n, (k+1)/n)$ and by crosses we denote points of type k/n:

Clearly

$$s_1 - s \le t - s, \quad t - t_1 \le t - s, \quad t_1 - s_1 \le t - s, \quad (t_1 - s_1)/n \le (t_1 - s_1)^2,$$

$$s_1 = ([ns]+1)/n, \quad t_1 = [nt]/n, \quad [nt] - ([ns]+1) = n(t_1 - s_1).$$

Hence and from (4) and the inequality $(a+b+c)^4 \le 81(a^4+b^4+c^4)$ we conclude that

$$E|\xi_t^n - \xi_s^n|^4 \le 81E(|\xi_t^n - \xi_{t_1}^n|^4 + |\xi_{t_1}^n - \xi_{s_1}^n|^4 + |\xi_{s_1}^n - \xi_s^n|^4)$$

$$\le 162(t-s)^2 m_4 + 81E|S_{[nt]}/\sqrt{n} - S_{[ns]+1}/\sqrt{n}|^4$$

$$= 162(t-s)^2 m_4 + 81n^{-2}a_{[nt]-([ns]+1)}$$

$$\leq 162(t-s)^2 m_4 + 243(t-s)^2 + 81(t_1 - s_1)m_4/n \leq 243(m_4+1)|t-s|^2.$$

Thus for all positions of s and t we have (3) with $N = 243(m_4+1)$. The lemma is proved.

Remember yet another definition from probability theory. We say that a sequence $\xi^n, n \geq 1$, of $\mathbb{R}^k$-valued random variables is *asymptotically normal* with parameters (m, R) if F_{ξ^n} converges weakly to the Gaussian distribution with parameters (m, R) (by F_ξ we denote the distribution of a random variable ξ). Below we use the fact that the weak convergence of distributions is equivalent to the pointwise convergence of their characteristic functions.

2. Lemma. *For every* $0 \leq t_1 < t_2 < ... < t_k \leq 1$ *the vectors* $(\xi^n_{t_1}, \xi^n_{t_2}, ..., \xi^n_{t_k})$ *are asymptotically normal with parameters* $(0, (t_i \wedge t_j))$.

Proof. We only consider the case $k = 2$. Other k's are treated similarly. We have

$$\lambda_1 \xi^n_{t_1} + \lambda_2 \xi^n_{t_2} = (\lambda_1 + \lambda_2) S_{[nt_1]}/\sqrt{n} + \lambda_2 (S_{[nt_2]} - S_{[nt_1]+1})/\sqrt{n}$$

$$+\eta_{[nt_1]+1}\{(nt_1 - [nt_1])\lambda_1/\sqrt{n} + \lambda_2/\sqrt{n}\} + \eta_{[nt_2]+1}(nt_2 - [nt_2])\lambda_2/\sqrt{n}.$$

On the right, we have a sum of independent terms. In addition, the coefficients of $\eta_{[nt_1]+1}$ and $\eta_{[nt_2]+1}$ go to zero and

$$E\exp(ia_n \eta_{[nt]+1}) = E\exp(ia_n \eta_1) \to 1 \quad \text{as} \quad a_n \to 0.$$

Finally, by the central limit theorem, for $\varphi(\lambda) = E\exp(i\lambda\eta_1)$,

$$\lim_{n\to\infty} \varphi^n(\lambda/\sqrt{n}) = e^{-\lambda^2/2}.$$

Hence,

$$\lim_{n\to\infty} Ee^{i(\lambda_1 \xi^n_{t_1} + \lambda_2 \xi^n_{t_2})} = \lim_{n\to\infty} \big(\varphi(\lambda_1/\sqrt{n} + \lambda_2/\sqrt{n})\big)^{[nt_1]} \big(\varphi(\lambda_2/\sqrt{n})\big)^{[nt_2]-[nt_1]-1}$$

$$= \exp\{-((\lambda_1+\lambda_2)^2 t_1 + \lambda_2^2(t_2 - t_1))/2\}$$

$$= \exp\{-(\lambda_1^2 (t_1 \wedge t_1) + 2\lambda_1\lambda_2(t_1 \wedge t_2) + \lambda_2^2(t_2 \wedge t_2))/2\}.$$

The lemma is proved.

3. Theorem (Donsker). *The sequence of distributions* F_{ξ^n} *weakly converges on* C *to a measure. This measure is called the Wiener measure.*

Proof. Owing to Lemma 1, there is a sequence $n_i \to \infty$ such that $F_{\xi^{n_i}_{\cdot}}$ converges weakly to a measure μ. By Exercise 1.2.10 it only remains to prove that the limit is independent of the choice of subsequences.

Let $F_{\xi^{m_i}_{\cdot}}$ be another weakly convergent subsequence and ν its limit. Fix $0 \le t_1 < t_2 < ... < t_k \le 1$ and define a continuous function on C by the formula $\pi(x_\cdot) = (x_{t_1}, ..., x_{t_k})$. By Lemma 2, considering π as a random element on $(C, \mathfrak{B}(C), \mu)$, for every bounded continuous $f(x^1, ..., x^k)$, we get

$$\int_{\mathbb{R}^k} f(x^1, ..., x^k)\, \mu\pi^{-1}(dx) = \int_C f(x_{t_1}, ..., x_{t_k})\, \mu(dx_\cdot)$$

$$= \lim_{i\to\infty} \int_C f(x_{t_1}, ..., x_{t_k})\, F_{\xi^{n_i}_{\cdot}}(dx_\cdot) = \lim_{i\to\infty} Ef(\xi^{n_i}_{t_1}, ..., \xi^{n_i}_{t_k}) = Ef(\zeta_1, ..., \zeta_k),$$

where $(\zeta_1, ..., \zeta_k)$ is a random vector normally distributed with parameters $(0, t_i \wedge t_j)$. One gets the same result considering m_i instead of n_i. By Theorem 1.2.4, we conclude that $\mu\pi^{-1} = \nu\pi^{-1}$. This means that for every Borel $B^{(k)} \subset \mathbb{R}^k$ the measures μ and ν coincide on the set $\{x_\cdot : (x_{t_1}, ..., x_{t_k}) \in B^{(k)}\}$. The collection of all such sets (with varying k, $t_1, ..., t_k$) is an algebra. By a result from measure theory, a measure on a σ-field is uniquely determined by its values on an algebra generating the σ-field. Thus $\mu = \nu$ on $\mathfrak{B}(C)$, and the theorem is proved.

Below we will need the conclusion of the last argument from the above proof, showing that there can be only one measure on $\mathfrak{B}(C)$ with given values on finite dimensional cylinder subsets of C.

4. Remark. Since Gaussian distributions are uniquely determined by their means and covariances, finite-dimensional distributions of Gaussian processes are uniquely determined by mean value and covariance functions. Hence, given a continuous Gaussian process ξ_t, its distribution on $(C, \mathfrak{B}(C))$ is uniquely determined by the functions m_t and $R(s,t)$.

5. Definition. By a Wiener process we mean a continuous Gaussian process on $[0,1]$ with $m_t = 0$ and $R(s,t) = s \wedge t$.

As follows from above, the distributions of all Wiener processes on $(C, \mathfrak{B}(C))$ coincide if the processes exist at all.

6. Exercise*. Prove that if w_t is a Wiener process on $[0,1]$ and c is a constant with $c \ge 1$, then cw_{t/c^2} is also a Wiener process on $[0,1]$. This property is called *self-similarity* of the Wiener process.

7. Theorem. *There exists a Wiener process, and its distribution on* $(C, \mathfrak{B}(C))$ *is the Wiener measure.*

Proof. Let μ be the Wiener measure. On the probability space $(C, \mathfrak{B}(C), \mu)$ define the process $w_t(x.) = x_t$. Then, for every $0 \le t_1 < ... < t_k \le 1$ and continuous bounded $f(x^1, ..., x^k)$, as in the proof of Donsker's theorem, we have

$$Ef(w_{t_1}, ..., w_{t_k}) = \int_C f(x_{t_1}, ..., x_{t_k})\, \mu(dx.)$$

$$= \lim_{n \to \infty} Ef(\xi^n_{t_1}, ..., \xi^n_{t_k}) = Ef(\zeta^1, ..., \zeta^k),$$

where ζ is a Gaussian vector with parameters $(0, (t_i \wedge t_j))$. Since f is arbitrary, we see that the distribution of $(w_{t_1}, ..., w_{t_k})$ and $(\zeta^1, ..., \zeta^k)$ coincide, and hence $(w_{t_1}, ..., w_{t_k})$ is Gaussian with parameters $(0, (t_i \wedge t_j))$. Thus, w_t is a Gaussian process, $Ew_{t_i} = 0$, and $R(t_i, t_j) = Ew_{t_i}w_{t_j} = E\zeta_i\zeta_j = t_i \wedge t_j$. The theorem is proved.

This theorem and the remark before it show that the limit in Donsker's theorem is independent of the distributions of the η_k as long as $E\eta_k = 0$ and $E\eta_k^2 = 1$. In this framework Donsker's theorem is called *the invariance principle* (although there is no more "invariance" in this theorem than in the central limit theorem).

2. Some properties of the Wiener process

First we prove two criteria for a process to be a Wiener process.

1. Theorem. *A continuous process on* $[0, 1]$ *is a Wiener process if and only if*

(i) $w_0 = 0$ *(a.s.),*

(ii) $w_t - w_s$ *is normal with parameters* $(0, |t - s|)$ *for every* $s, t \in [0, 1]$,

(iii) $w_{t_1}, w_{t_2} - w_{t_1}, ... w_{t_n} - w_{t_{n-1}}$ *are independent for every* $n \ge 2$ *and* $0 \le t_1 \le t_2 \le ... \le t_n \le 1$.

Proof. First assume that w_t is a Wiener process. We have $w_0 \sim N(0, 0)$, hence $w_0 = 0$ (a.s.). Next take $0 \le t_1 \le t_2 \le ... \le t_n \le 1$ and let

$$\xi_1 = w_{t_1}, \quad \xi_2 = w_{t_2} - w_{t_1}, ..., \xi_n = w_{t_n} - w_{t_{n-1}}.$$

The vector $\xi = (\xi_1, ..., \xi_n)$ is a linear transform of $(w_{t_1}, ..., w_{t_n})$. Therefore ξ is Gaussian. In particular ξ_i and, generally, $w_t - w_s$ are Gaussian. Obviously, $E\xi_i = 0$ and, for $i > j$,

$$E\xi_i\xi_j = E(w_{t_i} - w_{t_{i-1}})(w_{t_j} - w_{t_{j-1}}) = Ew_{t_i}w_{t_j} - Ew_{t_{i-1}}w_{t_j} - Ew_{t_i}w_{t_{j-1}}$$

$$+ Ew_{t_{i-1}}w_{t_{j-1}} = t_j - t_j - t_{j-1} + t_{j-1} = 0.$$

Similarly, the equality $Ew_t w_s = s \wedge t$ implies that $E|w_t - w_s|^2 = |t - s|$. Thus $w_t - w_s \sim N(0, |t - s|)$, and we have proved (ii). In addition $\xi_i \sim N(0, t_i - t_{i-1})$, $E\xi_i^2 = t_i - t_{i-1}$, and

$$E \exp\{i \sum_k \lambda_k \xi_k\} = \exp\{-\frac{1}{2} \sum_{k,r} \lambda_k \lambda_r \mathrm{cov}\,(\xi_k, \xi_r)\}$$

$$= \exp\{-\frac{1}{2} \sum_k \lambda_k^2 (t_k - t_{k-1})\} = \prod_k E \exp\{i\lambda_k \xi_k\}.$$

This proves (iii).

Conversely, let w_t be a continuous process satisfying (i) through (iii). Again take $0 \le t_1 \le t_2 \le ... \le t_n \le 1$ and the same ξ_i's. From (i) through (iii), it follows that $(\xi_1, ..., \xi_n)$ is a Gaussian vector. Since $(w_{t_1}, ..., w_{t_n})$ is a linear function of $(\xi_1, ..., \xi_n)$, $(w_{t_1}, ..., w_{t_n})$ is also a Gaussian vector; hence w_t is a Gaussian process. Finally, for every $t_1, t_2 \in [0, 1]$ satisfying $t_1 \le t_2$, we have

$$m_{t_1} = E\xi_1 = 0, \quad R(t_1, t_2) = R(t_2, t_1) = Ew_{t_1} w_{t_2} = E\xi_1(\xi_1 + \xi_2)$$
$$= E\xi_1^2 = t_1 = t_1 \wedge t_2.$$

The theorem is proved.

2. Theorem. *A continuous process on* $[0, 1]$ *is a Wiener process if and only if*

(i) $w_0 = 0$ *(a.s.),*

(ii) $w_t - w_s$ *is normal with parameters* $(0, |t - s|)$ *for every* $s, t \in [0, 1]$,

(iii) *for every* $n \ge 2$ *and* $0 \le t_1 \le t_2 \le ... \le t_n \le 1$, *the random variable* $w_{t_n} - w_{t_{n-1}}$ *is independent of* $w_{t_1}, w_{t_2}, ... w_{t_{n-1}}$.

Proof. It suffices to prove that properties (iii) of this and the previous theorems are equivalent under the condition that (i) and (ii) hold. We are going to use the notation from the previous proof. If (iii) of the present theorem holds, then

$$E \exp\{i \sum_{k=1}^{n} \lambda_k \xi_k\} = E \exp\{i\lambda_n \xi_n\} E \exp\{i \sum_{k=1}^{n-1} \lambda_k \xi_k\},$$

since $(\xi_1, ..., \xi_{n-1})$ is a function of $(w_{t_1}, ..., w_{t_{n-1}})$. By induction,

$$E \exp\{i \sum_{k=1}^{n} \lambda_k \xi_k\} = \prod_k E \exp\{i\lambda_k \xi_k\}.$$

This proves property (iii) of the previous theorem. Conversely if (iii) of the previous theorem holds, then one can carry out the same computation in the opposite direction and get that ξ_n is independent of $(\xi_1, ..., \xi_{n-1})$ and of $(w_{t_1}, ..., w_{t_{n-1}})$, since the latter is a function of the former. The theorem is proved.

3. Theorem (Bachelier). *For every $t \in (0,1]$ we have $\max_{s\le t} w_s \sim |w_t|$, which is to say that for every $x \ge 0$*

$$P\{\max_{s\le t} w_s \le x\} = \frac{2}{\sqrt{2\pi t}} \int_0^x e^{-y^2/(2t)}\, dy.$$

Proof. Take independent identically distributed random variables η_k so that $P(\eta_k = 1) = P(\eta_k = -1) = 1/2$, and define ξ^n_t by (1.2). First we want to find the distribution of

$$\zeta^n = \max_{[0,1]} \xi^n_t = n^{-1/2} \max_{k\le n} S_k.$$

Observe that, for each n, the sequence $(S_1, ..., S_n)$ takes its every particular value with the same probability 2^{-n}. In addition, for each integer $i > 0$, the number of sequences favorable for the events

$$\{\max_{k\le n} S_k \ge i, S_n < i\} \quad \text{and} \quad \{\max_{k\le n} S_k \ge i, S_n > i\} \tag{1}$$

is the same. One proves this by using the reflection principle; that is, one takes each sequence favorable for the first event, keeps it until the moment when it reaches the level i and then *reflects* its remaining part about this level. This implies equality of the probabilities of the events in (1). Furthermore, due to the fact that i is an integer, we have

$$\{\zeta^n \ge in^{-1/2},\ \xi^n_1 < in^{-1/2}\} = \{\max_{k\le n} S_k \ge i, S_n < i\}$$

and

$$\{\zeta^n \ge in^{-1/2},\ \xi^n_1 > in^{-1/2}\} = \{\max_{k\le n} S_k \ge i, S_n > i\}.$$

Hence,

$$P\{\zeta^n \ge in^{-1/2},\ \xi^n_1 < in^{-1/2}\} = P\{\zeta^n \ge in^{-1/2},\ \xi^n_1 > in^{-1/2}\}.$$

Moreover, obviously,

$$P\{\zeta^n \ge in^{-1/2},\ \xi^n_1 > in^{-1/2}\} = P\{\xi^n_1 > in^{-1/2}\},$$

$$P\{\zeta^n \geq in^{-1/2}\} = P\{\zeta^n \geq in^{-1/2},\ \xi_1^n > in^{-1/2}\}$$
$$+ P\{\zeta^n \geq in^{-1/2},\ \xi_1^n < in^{-1/2}\} + P\{\xi_1^n = in^{-1/2}\}.$$

It follows that

$$P\{\zeta^n \geq in^{-1/2}\} = 2P\{\xi_1^n > in^{-1/2}\} + P\{\xi_1^n = in^{-1/2}\} \tag{2}$$

for every integer $i > 0$. The last equality also obviously holds for $i = 0$. We see that for numbers a of type $in^{-1/2}$, where i is a nonnegative integer, we have

$$P\{\zeta^n \geq a\} = 2P\{\xi_1^n > a\} + P\{\xi_1^n = a\}. \tag{3}$$

Certainly, the last probability goes to zero as $n \to \infty$ since ξ_1^n is asymptotically normal with parameters $(0, 1)$. Also, keeping in mind Donsker's theorem, it is natural to think that

$$P\{\max_{s\leq 1} \xi_s^n \geq a\} \to P\{\max_{s\leq 1} w_s \geq a\}, \quad 2P\{\xi_1^n > a\} \to 2P\{w_1 > a\}.$$

Therefore, (3) naturally leads to the conclusion that

$$P\{\max_{s\leq 1} w_s \geq a\} = 2P\{w_1 > a\} = P\{|w_1| > a\} \quad \forall a \geq 0,$$

and this is our statement for $t = 1$.

To justify the above argument, notice that (2) implies that

$$P\{\zeta^n = in^{-1/2}\} = P\{\zeta^n \geq in^{-1/2}\} - P\{\zeta^n \geq (i+1)n^{-1/2}\}$$
$$= 2P\{\xi_1^n = (i+1)n^{-1/2}\} + P\{\xi_1^n = in^{-1/2}\} - P\{\xi_1^n = (i+1)n^{-1/2}\}$$
$$= P\{\xi_1^n = (i+1)n^{-1/2}\} + P\{\xi_1^n = in^{-1/2}\}, \quad i \geq 0.$$

Now for every bounded continuous function $f(x)$ which vanishes for $x < 0$ we get

$$Ef(\zeta^n) = \sum_{i=0}^{\infty} f(in^{-1/2})P\{\zeta^n = in^{-1/2}\} = Ef(\xi_1^n - n^{-1/2}) + Ef(\xi_1^n).$$

By Donsker's theorem and by the continuity of the function $x_\cdot \to \max_{[0,1]} x_t$ we have

$$Ef(\max_{[0,1]} w_t) = 2Ef(w_1) = Ef(|w_1|).$$

We have proved our statement for $t = 1$. For smaller t one uses Exercise 1.6, saying that cw_{s/c^2} is a Wiener process for $s \in [0,1]$ if $c \geq 1$. The theorem is proved.

4. Theorem (on the modulus of continuity). *Let w_t be a Wiener process on $[0,1]$, $1/2 > \varepsilon > 0$. Then for almost every ω there exists $n \geq 0$ such that for each $s, t \in [0,1]$ satisfying $|t-s| \leq 2^{-n}$, we have*

$$|w_t - w_s| \leq N|t-s|^{1/2-\varepsilon},$$

where N depends only on ε. In particular, $|w_t| = |w_t - w_0| \leq Nt^{1/2-\varepsilon}$ for $t \leq 2^{-n}$.

Proof. Take a number $\alpha > 2$ and denote $\beta = \alpha/2 - 1$. Let $\xi \sim N(0,1)$. Since $w_t - w_s \sim N(0, |t-s|)$, we have $w_t - w_s \sim \xi|t-s|^{1/2}$. Hence

$$E|w_t - w_s|^\alpha = |t-s|^{\alpha/2} E|\xi|^\alpha = N_1(\alpha)|t-s|^{1+\beta}.$$

Next, let

$$K_n(a) = \{x_\cdot \in C : |x_0| \leq 2^n, |x_t - x_s| \leq N(a)|t-s|^a \quad \forall |t-s| \leq 2^{-n}\}.$$

By Theorem 1.4.6, for $0 < a < \beta\alpha^{-1}$, we have

$$P\{w_\cdot \in \bigcup_{n=1}^\infty K_n(a)\} = 1.$$

Therefore, for almost every ω there exists $n \geq 0$ such that for all $s, t \in [0,1]$ satisfying $|t-s| \leq 2^{-n}$, we have $|w_t(\omega) - w_s(\omega)| \leq N(a)|t-s|^a$. It only remains to observe that we can take $a = 1/2 - \varepsilon$ if from the very beginning we take $\alpha > 1/\varepsilon$ (for instance $\alpha = 2/\varepsilon$). The theorem is proved.

5. Exercise. Prove that there exists a constant N such that for almost every ω there exists $n \geq 0$ such that for each $s, t \in [0,1]$ satisfying $|t-s| \leq 2^{-n}$, we have

$$|w_t - w_s| \leq N\sqrt{|t-s|(-\ln|t-s|)},$$

The result of Exercise 5 is not far from the best possible. P. Lévy proved that

$$\varlimsup_{\substack{0 \leq s < t \leq 1 \\ u = t-s \to 0}} \frac{|w_t - w_s|}{\sqrt{2u(-\ln u)}} = 1 \quad \text{(a.s.)}.$$

6. Theorem (on quadratic variation). *Let* $0 = t_{0n} \leq t_{1n} \leq ... \leq t_{k_n n} = 1$ *be a sequence of partitions of* $[0,1]$ *such that* $\max_i(t_{i+1,n} - t_{in}) \to 0$ *as* $n \to \infty$. *Also let* $0 \leq s \leq t \leq 1$. *Then, in probability as* $n \to \infty$,

$$\sum_{s \leq t_{in} \leq t_{i+1,n} \leq t} (w_{t_{i+1,n}} - w_{t_{in}})^2 \to t - s. \tag{4}$$

Proof. Let

$$\xi_n := \sum_{s \leq t_{in} \leq t_{i+1,n} \leq t} (w_{t_{i+1,n}} - w_{t_{in}})^2$$

and observe that ξ_n is a sum of independent random variables. Also use that if $\eta \sim N(0, \sigma^2)$, then $\eta = \sigma\zeta$, where $\zeta \sim N(0,1)$, and $\operatorname{Var}\eta^2 = \sigma^4 \operatorname{Var}\zeta^2$. Then, for $N := \operatorname{Var}\zeta$, we obtain

$$\operatorname{Var}\xi_n = \sum_{s \leq t_{in} \leq t_{i+1,n} \leq t} \operatorname{Var}[(w_{t_{i+1,n}} - w_{t_{in}})^2] = N \sum_{s \leq t_{in} \leq t_{i+1,n} \leq t} (t_{i+1,n} - t_{in})^2$$

$$\leq N \max_i (t_{i+1,n} - t_{in}) \sum_{0 \leq t_{in} \leq t_{i+1,n} \leq 1} (t_{i+1,n} - t_{in}) = N \max_i (t_{i+1,n} - t_{in}) \to 0.$$

In particular, $\xi_n - E\xi_n \to 0$ in probability. In addition,

$$E\xi_n = \sum_{s \leq t_{in} \leq t_{i+1,n} \leq t} (t_{i+1,n} - t_{in}) \to t - s.$$

Hence $\xi_n - (t-s) = \xi_n - E\xi_n + E\xi_n - (t-s) \to 0$ in probability, and the theorem is proved.

7. Exercise. Prove that if $t_{in} = i/2^n$, then the convergence in (4) holds almost surely.

8. Corollary. *It is not true that there exist functions* $\varepsilon(\omega)$ *and* $N(\omega)$ *such that with positive probability* $\varepsilon(\omega) > 0$, $N(\omega) < \infty$, *and*

$$|w_t(\omega) - w_s(\omega)| \leq N(\omega)|t-s|^{1/2+\varepsilon(\omega)}$$

whenever $t, s \in [0,1]$ *and* $|t - s| \leq \varepsilon(\omega)$.

Indeed, if $|w_t(\omega) - w_s(\omega)| \leq N(\omega)|t-s|^{1/2+\varepsilon(\omega)}$ for $|t-s|$ sufficiently small, then

$$\sum_i (w_{t_{i+1,n}}(\omega) - w_{t_{in}}(\omega))^2 \leq N^2 \sum_i (t_{i+1,n} - t_{in})^{1+2\varepsilon} \to 0.$$

9. Corollary. $P\{\operatorname{Var}_{[0,1]} w_t = \infty\} = 1$.

This follows from the fact that, owing to the continuity of w_t,

$$\sum_i (w_{t_{i+1,n}}(\omega) - w_{t_{in}}(\omega))^2 \leq \max_i |w_{t_{i+1,n}}(\omega) - w_{t_{in}}(\omega)| \operatorname{Var}_{[0,1]} w_t(\omega) \to 0$$

if $\operatorname{Var}_{[0,1]} w_t(\omega) < \infty$.

10. Exercise. Let w_t be a one-dimensional Wiener process. Find

$$P\{\max_{s\leq 1} w_s \geq b, w_1 \leq a\}.$$

The following exercise is a particular case of the Cameron-Martin theorem regarding the process $w_t - \int_0^t f_s\,ds$ with nonrandom f. Its extremely powerful generalization for random f is known as Girsanov's Theorem 6.8.8.

11. Exercise. Let w_t be a one-dimensional Wiener process on a probability space $(\Omega, \mathcal{F}, P)$. Prove that

$$Ee^{w_t - t/2} = 1.$$

Introduce a new measure by $Q(d\omega) = e^{w_1 - 1/2}P(d\omega)$. Prove that $(\Omega, \mathcal{F}, Q)$ is a probability space, and that $w_t - t$, $t \in [0, 1]$, is a Wiener process on $(\Omega, \mathcal{F}, Q)$.

12. Exercise. By using the results in Exercise 11 and the fact that the distributions on $(C, \mathfrak{B}(C))$ of Wiener processes coincide, show that

$$P\{\max_{s\leq 1}[w_s + s] \leq a\} = Ee^{w_1 - 1/2} I_{\max_{s\leq 1} w_s \leq a}.$$

Then by using the result in Exercise 10, compute the last expectation.

Unboundedness of the variation of Wiener trajectories makes it hard to justify the following argument. In real situations the variance of Brownian motion of pollen grains should depend on the water temperature. If the temperature is piecewise constant taking constant value on each interval of a partition $0 \leq t_1 < t_2 < ... < t_n = 1$, then the trajectory can be modeled by

$$\sum_{t_{i+1}\leq t} (w_{t_{i+1}} - w_{t_i})f_i + (w_t - w_{t_k})f_k,$$

where $k = \max\{i : t_i \leq t\}$ and the factor f_i reflects the dependence of the variance on temperature for $t \in [t_i, t_{i+1})$. The difficulty comes when one tries to pass from piecewise constant temperatures to continuously changing ones, because the sum should converge to an integral against w_t as we make partitions finer and finer. On the other hand, the integral against w_t is not defined since the variation of w_t is infinite for almost each ω. Yet there is a rather narrow class of functions f, namely functions of bounded variation, for which one can define the Riemann integral against w_t pathwise (see Theorem 3.22). For more general functions one defines the integral against w_t in the mean-square sense.

3. Integration against random orthogonal measures

The reader certainly knows the basics of the theory of L_p spaces, which can be found, for instance, in [**Du**] and which we only need for $p = 1$ and $p = 2$. Our approach to integration against random orthogonal measures requires a version of this theory which starts with introducing step functions using not all measurable sets but rather some collection of them. Actually, the version is quite parallel to the usual theory, and what follows below should be considered as just a reminder of the general scheme of the theory of L_p spaces.

Let X be a set, Π *some* family of subsets of X, $\mathfrak{A}$ a σ-algebra of subsets of X, and μ a measure on $(X, \mathfrak{A})$. Suppose that $\Pi \subset \mathfrak{A}$ and $\Pi_0 := \{\Delta \in \Pi : \mu(\Delta) < \infty\} \neq \emptyset$. Let $S(\Pi) = S(\Pi, \mu)$ denote the set of all *step functions*, that is, functions

$$\sum_{i=1}^{n} c_i I_{\Delta(i)}(x),$$

where c_i are complex numbers, $\Delta(i) \in \Pi_0$ (not Π!), $n < \infty$ is an integer. For $p \in [1, \infty)$, let $L_p(\Pi, \mu)$ denote the set of all $\mathfrak{A}^\mu$-measurable complex-valued functions f on X for each of which there exists a sequence $f_n \in S(\Pi)$ such that

$$\int_X |f - f_n|^p \, \mu(dx) \longrightarrow 0 \quad \text{as} \quad n \to \infty. \tag{1}$$

A sequence $f_n \in S(\Pi)$ that satisfies (1) will be called a *defining sequence* for f. From the convexity of $|t|^p$, we infer that $|a+b|^p \leq 2^{p-1}|a|^p + 2^{p-1}|b|^p$, $|f|^p \leq 2^{p-1}|f_n|^p + 2^{p-1}|f - f_n|^p$ and therefore, if $f \in L_p(\Pi, \mu)$, then

$$\|f\|_p := \Big(\int_X |f|^p \, \mu(dx)\Big)^{1/p} < \infty. \tag{2}$$

The expression $\|f\|_p$ is called the L_p *norm* of f. For $p = 2$ it is also useful to define the *scalar product* (f, g) of elements $f, g \in L_2(\Pi, \mu)$:

$$(f,g) := \int_X f\bar{g}\,\mu(dx). \tag{3}$$

This integral exists and is finite, since $|f\bar{g}| \le |f|^2 + |g|^2$. The expression $\|f-g\|_p$ defines a distance in $L_p(\Pi,\mu)$ between the elements $f,g \in L_p(\Pi,\mu)$. It is "almost" a metric on $L_p(\Pi,\mu)$, in the sense that, although the equality $\|f-g\|_p = 0$ implies that $f=g$ only almost everywhere with respect to μ, nevertheless $\|f-g\|_p = \|g-f\|_p$ and the *triangle inequality* holds:

$$\|f+g\|_p \le \|f\|_p + \|g\|_p.$$

If $f_n, f \in L_p(\Pi,\mu)$ and $\|f_n - f\|_p \to 0$ as $n\to\infty$, we will naturally say that f_n *converges to* f *in* $L_p(\Pi,\mu)$. If $\|f_n - f_m\|_p \to 0$ as $n,m\to\infty$, we will call f_n a *Cauchy sequence* in $L_p(\Pi,\mu)$. The following results are useful. For their proofs we refer the reader to [**Du**].

1. Theorem. (i) *If f_n is a Cauchy sequence in $L_p(\Pi,\mu)$, then there exists a subsequence $f_{n(k)}$ such that $f_{n(k)}$ has a limit μ-a.e. as $k\to\infty$.*

(ii) *$L_p(\Pi,\mu)$ is a linear space, that is, if a,b are complex numbers and $f,g \in L_p(\Pi,\mu)$, then $af+bg \in L_p(\Pi,\mu)$.*

(iii) *$L_p(\Pi,\mu)$ is a complete space, that is, for every Cauchy sequence $f_n \in L_p(\Pi,\mu)$, there exists an $\mathfrak{A}$-measurable function f for which* (1) *is true; in addition, every $\mathfrak{A}^\mu$-measurable function f that satisfies* (1) *for some sequence $f_n \in L_p(\Pi,\mu)$ is an element of $L_p(\Pi,\mu)$.*

2. Exercise*. Prove that if Π is a *σ-field*, then $L_p(\Pi,\mu)$ is simply the set of all Π^μ-measurable functions f that satisfy (2).

3. Exercise. Prove that if Π_0 consists of only one set Δ, then $L_p(\Pi,\mu)$ is the set of all functions μ-almost everywhere equal to a constant times the indicator of Δ.

4. Exercise. Prove that if $(X,\mathfrak{A},\mu) = ([0,1],\mathfrak{B}[0,1],\ell)$ and $\Pi = \{(0,t] : t\in(0,1)\}$, then $L_p(\Pi,\mu)$ is the space of all Lebesgue measurable functions summable to the pth power on $[0,1]$.

We now proceed to the main contents of this section. Let $(\Omega,\mathcal{F},P)$ be a probability space and suppose that to every $\Delta\in\Pi_0$ there is assigned a random variable $\zeta(\Delta) = \zeta(\omega,\Delta)$.

5. Definition. We say that ζ is a *random orthogonal measure with reference measure* μ if (a) $E|\zeta(\Delta)|^2 < \infty$ for every $\Delta\in\Pi_0$, (b) $E\,\zeta(\Delta_1)\bar{\zeta}(\Delta_2) = \mu(\Delta_1\cap\Delta_2)$ for all $\Delta_1,\Delta_2\in\Pi_0$.

6. Example. If $(X, \mathfrak{A}, \mu) = (\Omega, \mathcal{F}, P)$ and $\Pi = \mathfrak{A}$, then $\zeta(\Delta) := I_\Delta$ is a random orthogonal measure with reference measure μ. In this case, for each ω, ζ is just the Dirac measure concentrated at ω.

Generally, random orthogonal measures are not measures for each ω, because they need not even be defined on a σ-field. Actually, the situation is even more interesting, as the reader will see from Exercise 21.

7. Example. Let w_t be a Wiener process on $[0, 1]$ and

$$(X, \mathfrak{A}, \mu) = ([0, 1], \mathfrak{B}([0, 1]), \ell).$$

Let $\Pi = \{[0, t] : t \in (0, 1]\}$ and, for each $\Delta = [0, t] \in \Pi$, let $\zeta(\Delta) = w_t$. Then, for $\Delta_i = [0, t_i] \in \Pi$, we have

$$E\zeta(\Delta_1)\overline{\zeta(\Delta_2)} = Ew_{t_1}w_{t_2} = t_1 \wedge t_2 = \ell(\Delta_1 \cap \Delta_2),$$

which shows that ζ is a random orthogonal measure with reference measure ℓ.

8. Exercise*. Let τ_n be a sequence of independent random variables exponentially distributed with parameter 1. Define a sequence of random variables $\sigma_n = \tau_1 + ... + \tau_n$ and the corresponding counting process

$$\pi_t = \sum_{n=1}^{\infty} I_{[\sigma_n, \infty)}(t).$$

Observe that π_t is a function of locally bounded variation (at least for almost all ω), so that the usual integral against $d\pi_t$ is well defined: if f vanishes outside a finite interval, then

$$\int_0^\infty f(t)\, d\pi_t = \sum_{n=1}^{\infty} f(\sigma_n).$$

Prove that, for every bounded continuous real-valued function f given on $\mathbb{R}$ and having compact support and every $s \in \mathbb{R}$,

$$\varphi(s) := E\exp\{i\int_0^\infty f(s+t)\, d\pi_t\} = \exp(\int_0^\infty (e^{if(s+t)} - 1)\, dt).$$

Conclude from here that $\pi_t - \pi_s$ has Poisson distribution with parameter $|t - s|$. In particular, prove $E\pi_t = t$ and $E(\pi_t - t)^2 = t$. Also prove that π_t is a process with independent increments, that is, $\pi_{t_2} - \pi_{t_1}, ..., \pi_{t_{k+1}} - \pi_{t_k}$ are independent as long as the intervals $(t_j, t_{j+1}]$ are disjoint. The process π_t is called *a Poisson process with parameter 1*.

9. Example. Take the Poisson process π_t from Exercise 8. Denote $m_t = \pi_t - t$. If $0 \le s \le t$, then

$$Em_s m_t = Em_s^2 + Em_s(m_t - m_s) = Em_s^2 = s = s \wedge t.$$

Therefore, if in Example 7 we replace w_t with π_t, we again have a random orthogonal measure with reference measure ℓ.

We will always assume that ζ satisfies the assumptions of Definition 5. Note that by Exercise 2 we have $\zeta(\Delta) \in L_2(\mathcal{F}, P)$ for every $\Delta \in \Pi_0$. The word "orthogonal" in Definition 5 comes from the fact that if $\Delta_1 \cap \Delta_2 = \emptyset$, then $\zeta(\Delta_1) \perp \zeta(\Delta_2)$ in the Hilbert space $L_2(\mathcal{F}, P)$. The word "measure" is explained by the property that if $\Delta, \Delta_i \in \Pi_0$, the Δ_i's are pairwise disjoint, and $\Delta = \bigcup_i \Delta_i$, then $\zeta(\Delta) = \sum_i \zeta(\Delta_i)$, where the series converges in the mean-square sense. Indeed,

$$\lim_{n\to\infty} E|\zeta(\Delta) - \sum_{i\le n} \zeta(\Delta_i)|^2$$

$$= \lim_{n\to\infty} [E|\zeta(\Delta)|^2 + \sum_{i\le n} E|\zeta(\Delta_i)|^2 - 2\mathrm{Re} \sum_{i\le n} E\zeta(\Delta)\bar{\zeta}(\Delta_i)]$$

$$= \lim_{n\to\infty} [\mu(\Delta) + \sum_{i\le n} \mu(\Delta_i) - 2\sum_{i\le n} \mu(\Delta_i)] = 0.$$

Interestingly enough, our explanation of the word "measure" is void in Examples 7 and 9, since there is no $\Delta \in \Pi$ which is representable as a countable union of disjoint members of Π.

10. Lemma. *Let $\Delta_i, \Gamma_j \in \Pi_0$, and let c_i, d_j be complex numbers, $i = 1, ..., n$, $j = 1, ..., m$. Assume $\sum_{i\le n} c_i I_{\Delta_i} = \sum_{j\le m} d_j I_{\Gamma_j}$ (μ-a.e.). Then*

$$\sum_{i\le n} c_i \zeta(\Delta_i) = \sum_{j\le m} d_j \zeta(\Gamma_j) \quad (a.s.), \tag{4}$$

$$E|\sum_{i\le n} c_i \zeta(\Delta_i)|^2 = \int_X |\sum_{i\le n} c_i I_{\Delta_i}|^2 \, \mu(dx). \tag{5}$$

Proof. First we prove (5). We have

$$E|\sum_{i\le n} c_i\zeta(\Delta_i)|^2 = \sum_{i,j\le n} c_i \bar{c}_j E\zeta(\Delta_i)\bar{\zeta}(\Delta_j) = \sum_{i,j\le n} c_i\bar{c}_j \mu(\Delta_i \cap \Delta_j)$$

$$= \int_X \sum_{i,j\le n} c_i\bar{c}_j I_{\Delta_i} I_{\Delta_j} \, \mu(dx) = \int_X |\sum_{i\le n} c_i I_{\Delta_i}|^2 \, \mu(dx).$$

Hence,

$$E\Big|\sum_{i\le n} c_i\zeta(\Delta_i) - \sum_{j\le m} d_j\zeta(\Gamma_j)\Big|^2 = \int_X \Big|\sum_{i\le n} c_i I_{\Delta_i} - \sum_{j\le m} d_j I_{\Gamma_j}\Big|^2 \,\mu(dx) = 0.$$

The lemma is proved.

11. Remark. The first statement of the lemma looks quite surprising in the situation when μ is concentrated at only one point x_0. Then the equality $\sum_{i\le n} c_i I_{\Delta_i} = \sum_{j\le m} d_j I_{\Gamma_j}$ holds μ-almost everywhere if and only if

$$\sum_{i\le n} c_i I_{\Delta_i}(x_0) = \sum_{j\le m} d_j I_{\Gamma_j}(x_0),$$

and this may hold for very different $c_i, \Delta_i, d_j, \Gamma_j$. Yet each time (4) holds true.

Next, on $S(\Pi)$ define an operator I by the formula

$$I : \sum_{i\le n} c_i I_{\Delta_i} \to \sum_{i\le n} c_i\zeta(\Delta_i).$$

In the future we will always identify two elements of an L_p space which coincide almost everywhere. Under this stipulation, Lemma 10 shows that I is a well defined linear unitary operator from a subset $S(\Pi)$ of $L_2(\Pi,\mu)$ into $L_2(\mathcal{F}, P)$. In addition, by definition $S(\Pi)$ is dense in $L_2(\Pi,\mu)$ and every isometric operator is uniquely extendible from a dense subspace to the whole space. By this we mean the following result, which we suggest as an exercise.

12. Lemma. *Let B_1 and B_2 be Banach spaces and B_0 a linear subset of B_1. Let a linear isometric operator I be defined on B_0 with values in B_2 ($|Ib|_{B_2} = |b|_{B_1}$ for every $b \in B_0$). Then there exists a unique linear isometric operator $\tilde I : \bar B_0 \to B_2$ ($\bar B_0$ is the closure of B_0 in B_1) such that $\tilde I b = Ib$ for every $b \in B_0$.*

Combining the above arguments, we arrive at the following.

13. Theorem. *There exists a unique linear operator $I : L_2(\Pi,\mu) \to L_2(\mathcal{F}, P)$ such that*

(i) $I(\sum_{i\le n} c_i I_{\Delta_i}) = \sum_{i\le n} c_i\zeta(\Delta_i)$ (a.s.) for all finite n, $\Delta_i \in \Pi_0$ and complex c_i;

(ii) $E|If|^2 = \int_X |f|^2\,\mu(dx)$ for all $f \in L_2(\Pi,\mu)$.

For $f \in L_2(\Pi,\mu)$ we write

$$If = \int_X f(x)\,\zeta(dx)$$

and we call If *the stochastic integral of* f with respect to ζ. Observe that, by continuity of I, to find If it suffices to construct step functions f_n converging to f in the $L_2(\Pi, \mu)$ sense, and then

$$\int_X f(x)\,\zeta(dx) = \operatorname*{l.i.m.}_{n\to\infty} \int_X f_n(x)\,\zeta(dx).$$

The operator I preserves not only the norm but also the scalar product:

$$E \int_X f(x)\,\zeta(dx) \overline{\int_X g(x)\,\zeta(dx)} = \int_X f\bar{g}\,\mu(dx), \quad f, g \in L_2(\Pi, \mu). \qquad (6)$$

This follows after comparing the coefficients of the complex parameter λ in the equal (by Theorem 13) polynomials $E|I(f+\lambda g)|^2$ and $\int |f+\lambda g|^2\,\mu(dx)$.

14. Exercise. Take π_t from Example 9. Prove that for every Borel $f \in L_2(0,1)$ the stochastic integral of f against $\pi_t - t$ equals the usual integral; that is,

$$-\int_0^1 f(s)\,ds + \sum_{\sigma_n \le 1} f(\sigma_n).$$

15. Remark. If $E\zeta(\Delta) = 0$ for every $\Delta \in \Pi_0$, then for every $f \in L_2(\Pi, \mu)$, we have

$$E \int_X f\,\zeta(dx) = 0.$$

Indeed, for $f \in S(\Pi)$, this equality is verified directly; for arbitrary $f \in L_2(\Pi, \mu)$ it follows from the fact that, by Cauchy's inequality for $f_n \in S(\Pi)$,

$$|E \int_X f\,\zeta(dx)|^2 = |E \int_X (f - f_n)\,\zeta(dx)|^2$$

$$\le E| \int_X (f - f_n)\,\zeta(dx)|^2 = \int_X |f - f_n|^2\,\mu(dx).$$

We now proceed to the question as to when $L_p(\Pi, \mu)$ and $L_p(\mathfrak{A}, \mu)$ coincide, which is important in applications. Remember the following definitions.

16. Definition. Let X be a set, $\mathfrak{B}$ a family of subsets of X. Then $\mathfrak{B}$ is called a *π-system* if $A_1 \cap A_2 \in \mathfrak{B}$ for every A_1, $A_2 \in \mathfrak{B}$. It is called a *λ-system* if

(i) $X \in \mathfrak{B}$ and $A_2 \setminus A_1 \in \mathfrak{B}$ for every $A_1, A_2 \in \mathfrak{B}$ such that $A_1 \subset A_2$;

(ii) for every $A_1, A_2, ... \in \mathfrak{B}$ such that $A_i \cap A_j = \emptyset$ when $i \neq j$, $\bigcup_{n=1}^{\infty} A_n \in \mathfrak{B}$.

A typical example of λ-system is given by the collection of all subsets on which two given probability measures coincide.

17. Exercise*. Prove that if $\mathfrak{B}$ is both a λ-system and a π-system, then it is a σ-field.

A very important property of π- and λ-systems is stated as follows.

18. Lemma. *If Λ is a λ-system and Π is a π-system and $\Pi \subset \Lambda$, then $\sigma(\Pi) \subset \Lambda$.*

Proof. Let Λ_1 denote the smallest λ-system containing Π (Λ_1 is the intersection of all λ-systems containing Π). It suffices to prove that $\Lambda_1 \supset \sigma(\Pi)$. To do this, it suffices to prove, by Exercise 17, that Λ_1 is a π-system, that is, it contains the intersection of every two of its sets. For $B \in \Lambda_1$ let $\Lambda(B)$ denote the family of *all* $A \in \Lambda_1$ such that $A \cap B \in \Lambda_1$. Obviously, $\Lambda(B)$ is a λ-system. In addition, if $B \in \Pi$, then $\Lambda(B) \supset \Pi$ (since Π is a π-system). Consequently, if $B \in \Pi$, then by the definition of Λ_1, we have $\Lambda(B) \supset \Lambda_1$. But this means that $\Lambda(A) \supset \Pi$ for each $A \in \Lambda_1$, so that as before, $\Lambda(A) \supset \Lambda_1$ for each $A \in \Lambda_1$, that is, Λ_1 is a π-system. The lemma is proved.

19. Theorem. *Let $\mathfrak{A}_1 = \sigma(\Pi)$. Assume that Π is a π-system and that there exists a sequence $\Delta(1), \Delta(2), ... \in \Pi_0$ such that $\Delta(n) \subset \Delta(n+1)$, $X = \bigcup_n \Delta(n)$. Then $L_p(\Pi, \mu) = L_p(\mathfrak{A}_1, \mu)$.*

Proof. Let Σ denote the family of all subsets A of X such that

$$I_A I_{\Delta(n)} \in L_p(\Pi, \mu)$$

for every n. Observe that Σ is a λ-system. Indeed for instance, if $A_1, A_2, ... \in \Sigma$ are pairwise disjoint and $A = \bigcup_k A_k$, then

$$I_A I_{\Delta(n)} = \sum_k I_{A_k} I_{\Delta(n)},$$

where the series converges in $L_p(\Pi, \mu)$ since $\bigcup_{k \geq m} A_k \downarrow \emptyset$ as $m \to \infty$, $\mu(\Delta(n)) < \infty$, and

$$\int_X |\sum_{k\geq m} I_{A_k} I_{\Delta(n)}|^p \,\mu(dx) = \int_X \sum_{k\geq m} I_{A_k} I_{\Delta(n)} \,\mu(dx) = \mu\big(\Delta(n)\cap \bigcup_{k\geq m} A_k\big) \to 0$$

as $m \to \infty$.

Since $\Sigma \supset \Pi$, because Π is a π-system, it follows by Lemma 18 that $\Sigma \supset \mathfrak{A}_1$. Consequently, it follows from the definition of $L_p(\mathfrak{A}_1, \mu)$ that $I_{\Delta(n)} f \in L_p(\Pi, \mu)$ for $f \in L_p(\mathfrak{A}_1, \mu)$ and $n \geq 1$. Finally, a straightforward application of the dominated convergence theorem shows that $||I_{\Delta(n)} f - f||_p \to 0$ as $n \to \infty$. Hence $f \in L_p(\Pi, \mu)$ if $f \in L_p(\mathfrak{A}_1, \mu)$ and $L_p(\mathfrak{A}_1, \mu) \subset L_p(\Pi, \mu)$. Since the reverse inclusion is obvious, the theorem is proved.

It turns out that, under the conditions of Theorem 19, one can extend ζ from Π_0 to the larger set $\mathfrak{A}_0 := \sigma(\Pi) \cap \{\Gamma : \mu(\Gamma) < \infty\}$. Indeed, for $\Gamma \in \mathfrak{A}_0$ we have $I_\Gamma \in L_2(\Pi, \mu)$, so that the definition

$$\tilde{\zeta}(\Gamma) = \int_X I_\Gamma \,\zeta(dx)$$

makes sense. In addition, if $\Gamma_1, \Gamma_2 \in \mathfrak{A}_0$, then by (6)

$$E\tilde{\zeta}(\Gamma_1)\overline{\tilde{\zeta}(\Gamma_2)} = E \int_X I_{\Gamma_1} \,\zeta(dx) \overline{\int_X I_{\Gamma_2} \,\zeta(dx)}$$

$$= \int_X I_{\Gamma_1} I_{\Gamma_2} \,\mu(dx) = \mu(\Gamma_1 \cap \Gamma_2).$$

Since obviously $\zeta(\Delta) = \tilde{\zeta}(\Delta)$ (a.s.) for every $\Delta \in \Pi_0$, we have an extension indeed. In Sec. 7 we will see that sometimes one can extend ζ even to a larger set than $\mathfrak{A}_0$.

20. Exercise. Let $X \in \Pi_0$, and let Π be a π-system. Show that if $\tilde{\zeta}_1$ and $\tilde{\zeta}_2$ are two extensions of ζ to $\sigma(\Pi)$, then

$$\int_X f(x) \,\tilde{\zeta}_1(dx) = \int_X f(x) \,\tilde{\zeta}_2(dx)$$

(a.s.) for every $f \in L_2(\sigma(\Pi), \mu)$. In particular, $\tilde{\zeta}_1(\Gamma) = \tilde{\zeta}_2(\Gamma)$ (a.s.) for any $\Gamma \in \sigma(\Pi)$.

21. Exercise. Come back to Example 7. By what is said above there is an extension of ζ to $\mathfrak{B}([0,1])$. By using the independence of increments of w_t, prove that

$$E \exp(-\sum_n |\zeta((a_{n+1}, a_n])|) = 0,$$

where $a_n = 1/n$. Derive from here that for almost every ω the function $\zeta(\Gamma), \Gamma \in \mathfrak{B}([0,1])$, has unbounded variation and hence cannot be a measure.

Let us apply the above theory of stochastic integration to modeling Brownian motion when the temperature varies in time.

Take the objects introduced in Example 7. By Theorem 19 (and by Exercise 2), for every $f \in L_2(0,1)$ (where $L_2(0,1)$ is the usual L_2 space of square integrable functions on $(0,1)$) the stochastic integral $\int_X f(t)\,\zeta(dt)$ is well defined. Usually, one writes this integral as

$$\int_0^1 f(t)\,dw_t.$$

Observe that (by the continuity of the integral) if $f^n \to f$ in $L_2(0,1)$, then

$$\int_0^1 f^n(t)\,dw_t \to \int_0^1 f(t)\,dw_t$$

in the mean-square sense. In addition, if

$$f^n(t) = \sum_i f(t_{i+1,n}) I_{t_{in}<t\le t_{i+1,n}} = \sum_i f(t_{i+1,n})[I_{t\le t_{i+1,n}} - I_{t\le t_{in}}]$$

with $0 \le t_{in} \le t_{i+1,n} \le 1$, then (by definition and linearity)

$$\int_0^1 f(t)\,dw_t = \underset{n\to\infty}{\text{l.i.m.}} \int_0^1 f^n(t)\,dw_t = \underset{n\to\infty}{\text{l.i.m.}} \sum_i f(t_{i+1,n})(w_{t_{i+1,n}} - w_{t_{in}}). \tag{7}$$

Naturally, the integral

$$\int_0^t f(s)\,dw_s := \int_0^1 I_{s\le t} f(s)\,dw_s$$

gives us a representation of Brownian motion in the environment with changing temperature. However, for each individual t this integral is an element of $L_2(\mathcal{F},P)$ and thus is uniquely defined only up to sets of probability zero. For describing individual trajectories of Brownian motion we should take an appropriate representative of $\int_0^t f(s)\,dw_s$ for each $t \in [0,1]$. At this moment it is absolutely not clear whether this choice can be performed so that we will have continuous trajectories, which is crucial from the practical point of view. Much later (see Theorem 6.1.10) we will prove that one can indeed make the right choice even when f is a random function. The good news is that this issue can be easily settled at least for some functions f.

22. Theorem. *Let $t \in [0,1]$, and let f be absolutely continuous on $[0,t]$. Then*

$$\int_0^t f(s)\,dw_s = f(t)w_t - \int_0^t w_s f'(s)\,ds \quad (a.s.).$$

Proof. Define $t_{in} = ti/n$. Then the functions $f^n(s) := f(t_{in})$ for $s \in (t_{in}, t_{i+1,n}]$ converge to $f(s)$ uniformly on $[0,t]$ so that (cf. (7)) we have

$$\int_0^t f(s)\,dw_s = \int_0^1 I_{s\le t} f(s)\,dw_s = \operatorname*{l.i.m.}_{n\to\infty} \sum_{i\le n-1} f(t_{in})(w_{t_{i+1,n}} - w_{t_{in}})$$

$$= f(t)w_t - \operatorname*{l.i.m.}_{n\to\infty} \sum_{i\le n-1} w_{t_{i+1,n}}\big(f(t_{i+1,n}) - f(t_{in})\big)$$

(summation by parts), where the last sum is written as

$$\int_0^t w_{\kappa(s,n)} f'(s)\,ds \tag{8}$$

with $\kappa(n,s) = t_{i+1,n}$ for $s \in (t_{in}, t_{i+1,n}]$. By the continuity of w_s we have $w_{\kappa(s,n)} \to w_s$ uniformly on $[0,t]$, and by the dominated convergence theorem (f' is integrable) we see that (8) converges to $\int_0^t w_s f'(s)\,ds$ for every ω. It only remains to remember that the mean-square limit coincides (a.s.) with the pointwise limit if both exist. The theorem is proved.

23. Exercise*. Prove that if a real-valued $f \in L_2(0,1)$, then $\int_0^t f(s)\,dw_s$, $t \in [0,1]$, is a Gaussian process with zero mean and covariance

$$R(s,t) = \int_0^{s\wedge t} f^2(u)\,du = \Big(\int_0^s f^2(u)\,du\Big) \wedge \Big(\int_0^t f^2(u)\,du\Big).$$

The construction of the stochastic integral with respect to a random orthogonal measure is not specific to probability theory. We have considered the case in which $\zeta(\Delta) \in L_2(\mathcal{F}, P)$, where P is a *probability* measure. Our arguments could be repeated almost word for word for the case of an arbitrary measure. It would then turn out that the Fourier integral of L_2 functions is a particular case of integrals with respect to random orthogonal measure. In this connection we offer the reader the following exercise.

24. Exercise. Let Π be the set of all intervals $(a,b]$, where $a,b \in (-\infty,\infty)$, $a<b$. For $\Delta = (a,b] \in \Pi$, define a function $\zeta(\Delta) = \zeta(\omega,\Delta)$ on $(-\infty,\infty)$ by

$$\zeta(\Delta) = \frac{1}{i\omega}\big(e^{i\omega b} - e^{i\omega a}\big) = \int_\Delta e^{i\omega x}\,dx.$$

Define $L_p = L_p(\Pi, \ell) = L_p(\mathfrak{B}(\mathbb{R}), \ell)$. Prove, using a change of variable, that the number $\big(\zeta(\Delta_1), \zeta(\Delta_2)\big)$ equals its complex conjugate, that is, it is real, and that $\|\zeta(\Delta)\|_2^2 = c\,\ell(\Delta)$ for $\Delta_1, \Delta_2, \Delta \in \Pi$, where c is a constant independent of Δ. Use this and the observation that $\zeta(\Delta_1 \cup \Delta_2) = \zeta(\Delta_1) + \zeta(\Delta_2)$ if $\Delta_1, \Delta_2, \Delta_1 \cup \Delta_2 \in \Pi$, $\Delta_1 \cap \Delta_2 = \emptyset$, to deduce that in that case $\big(\zeta(\Delta_1), \zeta(\Delta_2)\big) = 0$. Using the fact that $\Delta_1 = (\Delta_1 \setminus \Delta_2) \cup (\Delta_1 \cap \Delta_2)$ and adding an interval between Δ_1, Δ_2 if they do not intersect, prove that $\big(\zeta(\Delta_1), \zeta(\Delta_2)\big) = c\,\ell(\Delta_1 \cap \Delta_2)$ for every $\Delta_1, \Delta_2 \in \Pi$ and, consequently, that we can construct an integral with respect to ζ, such that *Parseval's equality* holds for every $f \in L_2$:

$$c\|f\|_2^2 = \Big\|\int f\zeta(dx)\Big\|_2^2.$$

Keeping in mind that for $f \in S(\Pi)$, obviously,

$$\int f\zeta(dx) = \int_{-\infty}^{\infty} f(x)e^{i\omega x}\,dx \quad \text{(a.e.)},$$

generalize this equality to all $f \in L_2 \cap L_1$. Putting $f = \exp(-x^2)$ and using the characteristic function of the normal distribution, prove that $c = 2\pi$. Finally, use Fubini's theorem to prove that for $f \in L_1$ and $-\infty < a < b < \infty$, we have

$$\int_a^b \Big(\int_{-\infty}^{\infty} \bar{f}(\omega)e^{i\omega x}\,d\omega\Big)dx = \int_{-\infty}^{\infty} \frac{1}{i\omega}\big(e^{i\omega b} - e^{i\omega a}\big)\bar{f}(\omega)\,d\omega.$$

In other words, if $f \in L_1 \cap L_2$, then $\big(\zeta(\Delta), f\big) = c(I_\Delta, g)$, where

$$\bar{g}(x) = c^{-1}\int \bar{f}(\omega)\zeta(x, d\omega),$$

and (by definition) this leads to the *inversion formula for the Fourier transform*:

$$f(\omega) = \int g(x)\zeta(\omega, dx).$$

Generalize this formula from the case $f \in L_1 \cap L_2$ to all $f \in L_2$.

4. The Wiener process on $[0,\infty)$

The definition of the Wiener process on $[0,\infty)$ is the same as on $[0,1]$ (cf. Definition 1.5). Clearly for the Wiener process on $[0,\infty)$ one has the corresponding counterparts of Theorems 2.1 and 2.2 about the independence of increments and the independence of increments of previous values of the process. Also as in Exercise 1.6, if w_t is a Wiener process on $[0,\infty)$ and c is a strictly positive constant, then cw_{t/c^2} is also a Wiener process on $[0,\infty)$. This property is called *self-similarity* of the Wiener process.

1. Theorem. *There exists a Wiener process defined on $[0,\infty)$.*

Proof. Take any smooth function $f(t)>0$ on $[0,1)$ such that

$$\int_0^1 f^2(t)\,dt=\infty.$$

Let $\varphi(r)$ be the inverse function to $\int_0^t f^2(s)\,ds$. For $t<1$ define

$$y(t)=f(t)w_t-\int_0^t w_s f'(s)\,ds.$$

Obviously $y(t)$ is a continuous process. By Theorem 3.22 we have

$$y(t)=\int_0^t f(s)\,dw_s=\int_0^1 I_{s\le t}f(s)\,dw_s \quad \text{(a.s.)}.$$

By Exercise 3.23, y_t is a Gaussian process with zero mean and covariance

$$\int_0^{s\wedge t} f^2(u)\,du=\Big(\int_0^s f^2(u)\,du\Big)\wedge\Big(\int_0^t f^2(u)\,du\Big),\qquad s,t<1.$$

Now, as is easy to see, $x(r):=y(\varphi(r))$ is a continuous Gaussian process defined for $r\in[0,\infty)$ with zero mean and covariance $r_1\wedge r_2$. The theorem is proved.

Apart from the properties of the Wiener process on $[0,\infty)$ stated in the beginning of this section, which are similar to the properties on $[0,1]$, there are some new ones, of which we will state and prove only two.

2. Theorem. *Let w_t be a Wiener process for $t\in[0,\infty)$ defined on a probability space $(\Omega,\mathcal{F},P)$. Then there exists a set $\Omega'\in\mathcal{F}$ such that $P(\Omega')=1$ and, for each $\omega\in\Omega'$, we have*

$$\lim_{t\downarrow 0} tw_{1/t}(\omega)=0.$$

Furthermore, for $t>0$ define

$$\xi_t(\omega) = \begin{cases} tw_{1/t}(\omega) & \textit{if} \quad \omega \in \Omega', \\ 0 & \textit{if} \quad \omega \notin \Omega', \end{cases}$$

and let $\xi_0(\omega) \equiv 0$. *Then* ξ_t *is a Wiener process.*

Proof. Define $\tilde{\xi}_t = tw_{1/t}$ for $t > 0$ and $\tilde{\xi}_0 \equiv 0$. As is easy to see, $\tilde{\xi}_t$ is a Gaussian process with zero mean and covariance $s \wedge t$. It is also continuous on $(0,\infty)$. It follows, in particular, that $\sup_{s\in(0,t]} |\tilde{\xi}_s(\omega)|$ equals the sup over rational numbers on $(0,t]$. Since this sup is an increasing function of t, its limit as $t \downarrow 0$ can also be calculated along rational numbers. Thus,

$$\Omega' := \{\omega : \lim_{t\downarrow 0} \sup_{s\in(0,t]} |\tilde{\xi}_s(\omega)| = 0\} \in \mathcal{F}.$$

Next, let C' be the set of all (maybe unbounded) continuous functions on $(0,1]$, and $\Sigma(C')$ the cylinder σ-field of subsets of C', that is, the smallest σ-field containing all sets $\{x_\cdot \in C' : x_t \in \Gamma\}$ for all $t \in (0,1]$ and $\Gamma \in \mathfrak{B}(\mathbb{R})$. Then the distributions of $\tilde{\xi}_\cdot$ and $w_\cdot$ on $(C', \Sigma(C'))$ coincide (cf. Remark 1.4).

Define

$$A = \{x_\cdot \in C' : \lim_{t\downarrow 0} \sup_{s\in(0,t]} |x_s| = 0\}.$$

Since $x_\cdot \in C'$ are continuous in $(0,1]$, it is easy to see that $A \in \Sigma(C')$. Therefore,

$$P(\tilde{\xi}_\cdot \in A) = P(w_\cdot \in A),$$

which is to say,

$$P(\lim_{t\downarrow 0} \sup_{s\in(0,t]} |\tilde{\xi}_s| = 0) = P(\lim_{t\downarrow 0} \sup_{s\in(0,t]} |w_s| = 0).$$

The last probability being 1, we conclude that $P(\Omega') = 1$, and it only remains to observe that ξ_t is a continuous process and $\xi_t = \tilde{\xi}_t$ on Ω' or almost surely, so that ξ_t is a Gaussian process with zero mean and covariance $s \wedge t$. The theorem is proved.

3. Corollary. *Let* $1/2 > \varepsilon > 0$. *By Theorem* 2.4 *for almost every* ω *there exists* $n(\omega) < \infty$ *such that* $|\xi_t(\omega)| \le Nt^{1/2-\varepsilon}$ *for* $t \le 2^{-n(\omega)}$, *where* N *depends only on* ε. *Hence, for* w_t, *for almost every* ω *we have* $|w_t| \le Nt^{1/2+\varepsilon}$ *if* $t \ge 2^{n(\omega)}$.

4. Remark. Having the Wiener process on $[0,\infty)$, we can repeat the construction of the stochastic integral and define $\int_0^\infty f(t)\,dw_t$ for every $f \in L_2([0,\infty))$ starting with the random orthogonal measure $\zeta(0,a] = w_a$ defined for all $a \geq 0$. Of course, this integral has properties similar to those of $\int_0^1 f(t)\,dw_t$. In particular, the results of Theorem 3.22 on integrating by parts and of Exercise 3.23 still hold.

5. Markov and strong Markov properties of the Wiener process

Let $(\Omega, \mathcal{F}, P)$ be a probability space carrying a Wiener process w_t, $t \in [0,\infty)$. Also assume that for every $t \in [0,\infty)$ we are given a σ-field $\mathcal{F}_t \subset \mathcal{F}$ such that $\mathcal{F}_s \subset \mathcal{F}_t$ for $t \geq s$. We call such a collection of σ-fields an (increasing) *filtration of σ-fields*.

A trivial example of filtration is given by $\mathcal{F}_t \equiv \mathcal{F}$.

1. Definition. Let Σ be a σ-field, $\Sigma \subset \mathcal{F}$ and ξ a random variable taking values in a measurable space $(X, \mathfrak{B})$. We say that ξ and Σ are *independent* if $P(A, \xi \in B) = P(A)P(\xi \in B)$ for every $A \in \Sigma$ and $B \in \mathfrak{B}$.

2. Exercise*. Prove that if ξ and Σ are independent, $f(x)$ is a measurable function, and η is Σ-measurable, then $f(\xi)$ and η are independent as well.

3. Definition. We say that w_t is a Wiener process relative to the filtration $\mathcal{F}_t$ if w_t is $\mathcal{F}_t$-measurable for every t and $w_{t+h} - w_t$ is independent of $\mathcal{F}_t$ for every $t, h \geq 0$. In that case the couple $(w_t, \mathcal{F}_t)$ is called *a Wiener process*.

Below we assume that $(w_t, \mathcal{F}_t)$ is a Wiener process, explaining first that there always exists a filtration with respect to which w_t is a Wiener process.

4. Lemma. *Let*

$$\mathcal{F}_t^w := \sigma\{\{\omega : w_s(\omega) \in B\}, s \leq t, B \in \mathfrak{B}(\mathbb{R})\}.$$

Then $(w_t, \mathcal{F}_t^w)$ is a Wiener process.

Proof. By definition $\mathcal{F}_t^w$ is the smallest σ-field containing all sets $\{\omega : w_s(\omega) \in B\}$ for $s \leq t$ and Borel B. Since each of them is (as an element) in $\mathcal{F}$, $\mathcal{F}_t^w \subset \mathcal{F}$. The inclusion $\mathcal{F}_s^w \subset \mathcal{F}_t^w$ for $t \geq s$ is obvious, since $\{\omega : w_r(\omega) \in B\}$ belong to $\mathcal{F}_t^w$ for $r \leq s$ and $\mathcal{F}_s^w$ is the smallest σ-field containing them. Therefore $\mathcal{F}_t^w$ is a filtration.

Next, $\{\omega : w_t(\omega) \in B\} \in \mathcal{F}_t^w$ for $B \in \mathfrak{B}(\mathbb{R})$; hence w_t is $\mathcal{F}_t^w$-measurable. To prove the independence of $w_{t+h} - w_t$ and $\mathcal{F}_t^w$, fix a $B \in \mathfrak{B}(\mathbb{R})$, $t, h \geq 0$, and define

$$\mu(A) = P(A, w_{t+h} - w_t \in B), \quad \nu(A) = P(A)P(w_{t+h} - w_t \in B).$$

One knows that μ and ν are measures on $(\Omega, \mathcal{F})$. By Theorem 2.2 these measures coincide on every A of type $\{\omega : (w_{t_1}(\omega), ..., w_{t_n}(\omega)) \in B^{(n)}\}$ provided that $t_i \leq t$ and $B^{(n)} \in \mathfrak{B}(\mathbb{R}^n)$. The collection of these sets is an algebra (Exercise 1.3.3). Therefore μ and ν coincide on the smallest σ-field, say Σ, containing these sets. Observe that $\mathcal{F}_t^w \subset \Sigma$, since the collection generating Σ contains $\{\omega : w_s(\omega) \in D\}$ for $s \leq t$ and $D \in \mathfrak{B}(\mathbb{R})$. Hence μ and ν coincide on $\mathcal{F}_t^w$. It only remains to remember that B is an arbitrary element of $\mathfrak{B}(\mathbb{R})$. The lemma is proved.

We see that one can always take $\mathcal{F}_t^w$ as $\mathcal{F}_t$. However, it turns out that sometimes it is very inconvenient to restrict our choice of $\mathcal{F}_t$ to $\mathcal{F}_t^w$. For instance, we can be given a multi-dimensional Wiener process $(w_t^1, ..., w_t^d)$ (see Definition 6.4.1) and study only its first coordinate. In particular, while introducing stochastic integrals of random processes against dw_t^1 we may be interested in integrating functions depending not only on w_t^1 but on all other components as well.

5. Exercise*. Let $\bar{\mathcal{F}}_t^w$ be the completion of $\mathcal{F}_t^w$. Prove that $(w_t, \bar{\mathcal{F}}_t^w)$ is a Wiener process.

6. Theorem (Markov property). *Let $(w_t, \mathcal{F}_t)$ be a Wiener process. Fix t, $h_1, .., h_n \geq 0$. Then the vector $(w_{t+h_1} - w_t, ..., w_{t+h_n} - w_t)$ and the σ-field $\mathcal{F}_t$ are independent. Furthermore, $w_{t+s} - w_t$, $s \geq 0$, is a Wiener process.*

Proof. The last statement follows directly from the definitions. To prove the first one, without losing generality we assume that $h_1 \leq ... \leq h_n$ and notice that, since $(w_{t+h_1} - w_t, ..., w_{t+h_n} - w_t)$ is obtained by a linear transformation from η_n, where $\eta_k = (w_{t+h_1} - w_{t+h_0}, ..., w_{t+h_k} - w_{t+h_{k-1}})$ and $h_0 = 0$, we need only show that η_n and $\mathcal{F}_t$ are independent. We are going to use the theory of characteristic functions. Take $A \in \mathcal{F}_t$ and a vector $\lambda \in \mathbb{R}^n$. Notice that

$$EI_A \exp(i\lambda \cdot \eta_n) = EI_A \exp(i\mu \cdot \eta_{n-1}) \exp(i\lambda^n (w_{t+h_n} - w_{t+h_{n-1}})),$$

where $\mu = (\lambda^1, ..., \lambda^{n-1})$. Here I_A is $\mathcal{F}_t$-measurable and, since $\mathcal{F}_t \subset \mathcal{F}_{t+h_{n-1}}$, it is $\mathcal{F}_{t+h_{n-1}}$-measurable as well. It follows that $I_A \exp(i\mu \cdot \eta_{n-1})$ is $\mathcal{F}_{t+h_{n-1}}$-measurable. Furthermore, $w_{t+h_n} - w_{t+h_{n-1}}$ is independent of $\mathcal{F}_{t+h_{n-1}}$. Hence, by Exercise 2

$$EI_A \exp(i\lambda \cdot \eta_n) = EI_A \exp(i\mu \cdot \eta_{n-1}) E \exp(i\lambda^n (w_{t+h_n} - w_{t+h_{n-1}})),$$

and by induction and independence of increments of w_t

$$EI_A \exp(i\lambda \cdot \eta_n) = EI_A \prod_{j=1}^{n} E \exp(i\lambda^n (w_{t+h_j} - w_{t+h_{j-1}})) = P(A) E \exp(i\lambda \cdot \eta_n).$$

It follows from the theory of characteristic functions that for every Borel bounded g

$$EI_A g(\eta_n) = P(A)Eg(\eta_n).$$

It only remains to substitute here the indicator of a Borel set in place of g. The theorem is proved.

Theorem 6 says that, for every fixed $t \geq 0$, the process $w_{t+s} - w_t$, $s \geq 0$, starts afresh as a Wiener process forgetting everything that happened to w_r before time t. This property is quite natural for Brownian motion. It also has a natural extension when t is replaced with a random time τ, provided that τ does not depend on the future in a certain sense. To describe exactly what we mean by this, we need the following.

7. Definition. Let τ be a random variable taking values in $[0, \infty]$ (including ∞). We say that τ is a *stopping time* (relative to $\mathcal{F}_t$) if $\{\omega : \tau(\omega) > t\} \in \mathcal{F}_t$ for every $t \in [0, \infty)$.

The term "stopping time" is discussed after Exercise 3.3.3. Trivial examples of stopping times are given by nonrandom positive constants. A much more useful example is the following.

8. Example. Fix $a \geq 0$ and define

$$\tau = \tau_a = \inf\{t \geq 0 : w_t \geq a\} \quad (\inf \emptyset := \infty)$$

as the first hitting time of the point a by w_t. It turns out that τ is a stopping time.

Indeed, one can easily see that

$$\{\omega : \tau(\omega) > t\} = \{\omega : \max_{s \leq t} w_s(\omega) < a\}, \tag{1}$$

where, for ρ defined as the set of all rational points on $[0, \infty)$,

$$\max_{s \leq t} w_s = \sup_{r \in \rho, r \leq t} w_r,$$

which shows that $\max_{s \leq t} w_s$ is an $\mathcal{F}_t$-measurable random variable.

9. Exercise*. Let $a < 0 < b$ and let τ be the first exit time of w_t from (a, b):

$$\tau = \inf\{t \geq 0 : w_t \notin (a, b)\}.$$

Prove that τ is a stopping time.

10. Definition. Random processes $\eta_t^1, ..., \eta_t^n$ defined for $t \geq 0$ are called *independent* if for every $t_1, ..., t_k \geq 0$ the vectors $(\eta_{t_1}^1, ..., \eta_{t_k}^1), ..., (\eta_{t_1}^n, ..., \eta_{t_k}^n)$ are independent.

In what follows we consider some processes at random times, and these times occasionally can be infinite even though this happens with probability zero. In such situations we use the notation

$$x_\tau = x_\tau(\omega) = \begin{cases} x_{\tau(\omega)}(\omega) & \text{if} \quad \tau(\omega) < \infty, \\ 0 & \text{if} \quad \tau(\omega) = \infty. \end{cases}$$

11. Lemma. *Let $(w_t, \mathcal{F}_t)$ be a Wiener process and let τ be an $\mathcal{F}_t$-stopping time. Assume $P(\tau < \infty) = 1$. Then the processes $w_{t\wedge\tau}$ and $B_t := w_{\tau+t} - w_\tau$ are independent and the latter one is a Wiener process.*

Proof. Take $0 \leq t_1 \leq ... \leq t_k$. As is easy to see, we need only prove that for any Borel nonnegative functions $f(x_1, ..., x_k)$ and $g(x_1, ..., x_k)$

$$I_\tau := Ef(w_{t_1\wedge\tau}, ..., w_{t_k\wedge\tau})g(B_{t_1}, ..., B_{t_k})$$

$$= Ef(w_{t_1\wedge\tau}, ..., w_{t_k\wedge\tau})Eg(w_{t_1}, ..., w_{t_k}). \tag{2}$$

Assume for a moment that the set of values of τ is countable, say $r_1 < r_2 < ....$ By noticing that $\{\tau = r_n\} = \{\tau > r_{n-1}\} \setminus \{\tau > r_n\} \in \mathcal{F}_{r_n}$ and

$$F_n := f(w_{t_1\wedge\tau}, ..., w_{t_k\wedge\tau})I_{\tau=r_n} = f(w_{t_1\wedge r_n}, ..., w_{t_k\wedge r_n})I_{\tau=r_n},$$

we see that the first term is $\mathcal{F}_{r_n}$-measurable. Furthermore,

$$I_{\tau=r_n}g(B_{t_1}, ..., B_{t_k}) = I_{\tau=r_n}g(w_{r_n+t_1} - w_{r_n}, ..., w_{r_n+t_k} - w_{r_n}),$$

where, by Theorem 6, the last factor is independent of $\mathcal{F}_{r_n}$, and

$$Eg(w_{r_n+t_1} - w_{r_n}, ..., w_{r_n+t_k} - w_{r_n}) = Eg(w_{t_1}, ..., w_{t_k}).$$

Therefore,

$$I_\tau = \sum_{r_n} EF_n g(w_{r_n+t_1} - w_{r_n}, ..., w_{r_n+t_k} - w_{r_n})$$

$$= Eg(w_{t_1}, ..., w_{t_k}) \sum_{r_n} EF_n.$$

The last sum equals the first term on the right in (2). This proves the theorem for our particular τ.

In the general case we approximate τ and first notice (see, for instance, Theorem 1.2.4) that equation (2) holds for all Borel nonnegative f, g if and only if it holds for all bounded continuous f, g. Therefore, we assume f, g to be bounded and continuous.

Now, for $n = 1, 2, ...$, define

$$\tau_n(\omega) = (k+1)2^{-n} \quad \text{for } \omega \text{ such that} \quad k2^{-n} < \tau(\omega) \leq (k+1)2^{-n}, \qquad (3)$$

$k = -1, 0, 1, ...$ It is easily seen that $\tau \leq \tau_n \leq \tau + 2^{-n}$, $\tau_n \downarrow \tau$, and for $t \geq 0$

$$\{\omega : \tau_n > t\} = \{\omega : \tau(\omega) > 2^{-n}[2^n t]\} \in \mathcal{F}_{2^{-n}[2^n t]} \subset \mathcal{F}_t,$$

so that τ_n are stopping times. Hence, by the above result,

$$I_\tau = \lim_{n\to\infty} I_{\tau_n} = Eg(w_{t_1}, ..., w_{t_k}) \lim_{n\to\infty} Ef(w_{t_1\wedge\tau_n}, ..., w_{t_k\wedge\tau_n}),$$

and this leads to (2). The lemma is proved.

The following theorem states that the Wiener process has the strong Markov property.

12. Theorem. *Let $(w_t, \mathcal{F}_t)$ be a Wiener process and τ an $\mathcal{F}_t$-stopping time. Assume that $P(\tau < \infty) = 1$. Let*

$$\mathcal{F}^w_{\leq\tau} = \sigma\{\{\omega : w_{s\wedge\tau} \in B\}, s \geq 0, B \in \mathfrak{B}(\mathbb{R})\},$$

$$\mathcal{F}^w_{\geq\tau} = \sigma\{\{\omega : w_{\tau+s} - w_\tau \in B\}, s \geq 0, B \in \mathfrak{B}(\mathbb{R})\}.$$

Then the σ-fields $\mathcal{F}^w_{\leq\tau}$ and $\mathcal{F}^w_{\geq\tau}$ are independent in the sense that for every $A \in \mathcal{F}^w_{\leq\tau}$ and $B \in \mathcal{F}^w_{\geq\tau}$ we have $P(AB) = P(A)P(B)$. Furthermore, $w_{\tau+t} - w_\tau$ is a Wiener process.

Proof. The last assertion is proved in Lemma 11. To prove the first one we follow the proof of Lemma 4 and first let $B = \{\omega : (w_{\tau+s_1} - w_\tau, ..., w_{\tau+s_k} - w_\tau) \in \Gamma\}$, where $\Gamma \in \mathfrak{B}(\mathbb{R}^k)$. Consider two measures $\mu(A) = P(AB)$ and $\nu(A) = P(A)P(B)$ as measures on sets A. By Lemma 11 these measures coincide on every A of type $\{\omega : (w_{t_1\wedge\tau}, ..., w_{t_n\wedge\tau}) \in B^{(n)}\}$ provided that $B^{(n)} \in \mathfrak{B}(\mathbb{R}^n)$. The collection of these sets is an algebra (Exercise 1.3.3). Therefore μ and ν coincide on the smallest σ-field, which is $\mathcal{F}^w_{\leq\tau}$, containing these sets. Hence $P(AB) = P(A)P(B)$ for all $A \in \mathcal{F}^w_{\leq\tau}$ and our particular B. It only remains to repeat this argument relative to B upon fixing A. The theorem is proved.

6. Examples of applying the strong Markov property

First, we want to apply Theorem 5.12 to τ_a from Example 5.8. Notice that Bachelier's Theorem 2.3 holds not only for $t \in (0,1]$ but for $t \geq 1$ as well. One proves this by using the self-similarity of the Wiener process (cw_{t/c^2} is a Wiener process for every constant $c \neq 0$). Then, owing to (5.1), for $t > 0$ we find that $P(\tau_a > t) = P(|w_t| < a) = P(|w_1|\sqrt{t} < a)$, which tends to zero as $t \to \infty$, showing that $P(\tau_a < \infty) = 1$. Now Theorem 5.12 allows us to conclude that $w_{\tau+t} - w_\tau = w_{\tau+t} - a$ is a Wiener process independent of the trajectory on $[0, \tau]$. This makes rigorous what is quite clear intuitively. Namely, after reaching a, the Wiener process starts "afresh", forgetting everything which happened to it before. The same happens when it reaches a higher level $b > a$ after reaching a, and moreover, $\tau_b - \tau_a$ has the same distribution as τ_{b-a}. This is part of the following theorem, in which, as well as above, we allow ourselves to consider random variables like $\tau_b - \tau_a$ which may not be defined on a set of probability zero. We set $\tau_b(\omega) - \tau_a(\omega) = 0$ if $b > a > 0$ and $\tau_b(\omega) = \tau_a(\omega) = \infty$.

1. Theorem. (i) *For every* $0 < a_1 < a_2 < ... < a_n < \infty$ *the random variables* $\tau_{a_1}, \tau_{a_2} - \tau_{a_1}, ..., \tau_{a_n} - \tau_{a_{n-1}}$ *are independent.*

(ii) *For* $0 < a < b$, *the law of* $\tau_b - \tau_a$ *coincides with that of* τ_{b-a}, *and* τ_a *has* Wald's distribution *with density*

$$p(t) = (2\pi)^{-1/2} a t^{-3/2} \exp(-a^2/(2t)), \quad t > 0.$$

Proof. (i) It suffices to prove that $\tau_{a_n} - \tau_{a_{n-1}}$ is independent of $\tau_{a_1}, ..., \tau_{a_{n-1}}$ (cf. the proof of Theorem 2.2). To simplify notation, put $\tau(a) = \tau_a$. Since $a_i \leq a_{n-1}$ for $i \leq n-1$, we can rewrite (5.1) as

$$\{\omega : \tau(a_i) > t\} = \{\omega : \sup_{s \in \rho, s \leq t} w_{s \wedge \tau(a_{n-1})} < a_i\},$$

which implies that the $\tau(a_i)$ are $\mathcal{F}_{\leq \tau(a_{n-1})}$-measurable. On the other hand, for $t \geq 0$,

$$\{\omega : \tau(a_n) - \tau(a_{n-1}) > t\}$$

$$= \{\omega : \tau(a_n) - \tau(a_{n-1}) > t, \tau(a_{n-1}) < \infty\}$$

$$= \{\omega : \sup_{s \in \rho, s \leq t} (w_{\tau(a_{n-1})+s} - w_{\tau(a_{n-1})}) < a_n - a_{n-1}, \tau(a_{n-1}) < \infty\}$$

$$= \{\omega : 0 < \sup_{s \in \rho, s \leq t} (w_{\tau(a_{n-1})+s} - w_{\tau(a_{n-1})}) < a_n - a_{n-1}\}, \tag{1}$$

which shows that $\tau(a_n) - \tau(a_{n-1})$ is $\mathcal{F}_{\geq \tau(a_{n-1})}$-measurable. Referring to Theorem 5.12 finishes the proof of (i).

(ii) Let $n = 2$, $a_1 = a$, and $a_2 = b$. Then in the above notation $\tau(a_n) = \tau_b$ and $\tau(a_{n-1}) = \tau_a$. Since $w_{\tau(a_{n-1})+t} - w_{\tau(a_{n-1})} = w_{\tau_a+t} - w_{\tau_a}$ is a Wiener process and the distributions of Wiener processes coincide, the probability of the event on the right in (1) equals

$$P(\sup_{s\in\rho, s\le t} w_s < a_n - a_{n-1} = b - a) = P(\tau_{b-a} > t).$$

This proves the first assertion in (ii). To find the distribution of τ_a, remember that

$$P(\tau_a > t) = P(\max_{s\le t} w_s < a) = P(|w_1|\sqrt{t} < a) = \frac{2}{\sqrt{2\pi}} \int_0^{a/\sqrt{t}} e^{-y^2/2}\, dy.$$

By differentiating this formula we immediately get our density. The theorem is proved.

2. Exercise. We know that the Wiener process is self-similar in the sense that cw_{t/c^2} is a Wiener process for every constant $c \ne 0$. The process τ_a, $a \ge 0$, also has this kind of property. Prove that, for every $c > 0$, the process $c\tau_{a/\sqrt{c}}$, $a \ge 0$, has the same finite-dimensional distributions as τ_a, $a \ge 0$. Such processes are called *stable*. The Wiener process is a stable process of order 2, and the process τ_a is a stable process of order 1/2.

Our second application exhibits the importance of the operator $u \to u''$ in computing various expectations related to the Wiener process. The following results can be obtained quite easily on the basis of Itô's formula from Chapter 6. However, the reader might find it instructive to see that there is a different approach using the strong Markov property.

3. Lemma. *Let u be a twice continuously differentiable function defined on $\mathbb{R}$ such that u, u', and u'' are bounded. Then, for every $\lambda > 0$,*

$$u(0) = E\int_0^\infty e^{-\lambda t}(\lambda u(w_t) - (1/2)u''(w_t))\, dt. \tag{2}$$

Proof. Since w_t is a normal $(0, t)$ variable, the right-hand side of (2) equals

$$I := \int_0^\infty e^{-\lambda t} E(\lambda u(w_t) - (1/2)u''(w_t))\, dt$$

$$= \int_0^\infty e^{-\lambda t}\Big(\int_{\mathbb{R}} (\lambda u(x) - (1/2)u''(x))p(t,x)\, dx\Big)\, dt,$$

where

$$p(t,x) := \frac{1}{\sqrt{2\pi t}} e^{-x^2/(2t)}, \quad t > 0.$$

We continue our computation, integrating by parts. One can easily check that

$$\frac{1}{2}\frac{\partial^2 p}{(\partial x)^2} = \frac{\partial p}{\partial t}, \quad e^{-\lambda t}\lambda p - \frac{e^{-\lambda t}}{2}\frac{\partial^2 p}{(\partial x)^2} = -\frac{\partial}{\partial t}(e^{-\lambda t}p).$$

Hence

$$I = \lim_{\varepsilon \downarrow 0} \int_\varepsilon^\infty e^{-\lambda t}\Big(\int_{\mathbb{R}} (\lambda u(x) - (1/2)u''(x))p(t,x)\,dx\Big)\,dt$$

$$= -\lim_{\varepsilon \downarrow 0} \int_\varepsilon^\infty \frac{\partial}{\partial t}\Big(e^{-\lambda t}\int_{\mathbb{R}} u(x)p(t,x)\,dx\Big)\,dt = \lim_{\varepsilon \downarrow 0} e^{-\lambda\varepsilon}\int_{\mathbb{R}} u(x)p(\varepsilon,x)\,dx$$

$$= \lim_{\varepsilon \downarrow 0} Eu(w_\varepsilon) = u(0).$$

The lemma is proved.

4. Theorem. *Let $-\infty < a < 0 < b < \infty$, and let u be a twice continuously differentiable function given on $[a,b]$. Let τ be the first exit time of w_t from the interval (a,b) (see Exercise 5.9). Then, for every $\lambda \geq 0$,*

$$u(0) = E\int_0^\tau e^{-\lambda t}(\lambda u(w_t) - (1/2)u''(w_t))\,dt + Ee^{-\lambda\tau}u(w_\tau). \tag{3}$$

Proof. If needed, one can continue u outside $[a,b]$ and have a function, for which we keep the same notation, satisfying the assumptions of Lemma 3. Denote $f = \lambda u - u''$. Notice that obviously $\tau \leq \tau_b$, and, as we have seen above, $P(\tau_b < \infty) = 1$. Therefore by Lemma 3 we find that, for $\lambda > 0$,

$$u(0) = E\int_0^\infty ... = E\int_0^\tau ... + E\int_\tau^\infty ...$$

$$= E\int_0^\tau e^{-\lambda t}f(w_t)\,dt + \int_0^\infty e^{-\lambda t}Ee^{-\lambda\tau}f(w_\tau + B_t)\,dt =: I + J,$$

where $B_t = w_{\tau+t} - w_\tau$. Now we want to use Theorem 5.12. The reader who did Exercise 5.9 understands that τ is $\mathcal{F}^w_{\leq\tau}$-measurable. Furthermore, $w_{t\wedge\tau}I_{\tau<\infty} \to w_\tau$ as $t \to \infty$, so that w_τ is also $\mathcal{F}^w_{\leq\tau}$-measurable. Hence (τ, w_τ) and B_t are independent, and

$$J = \int_0^\infty e^{-\lambda t}Ee^{-\lambda\tau}f(w_\tau + B_t)\,dt = Ee^{-\lambda\tau}v(w_\tau),$$

where

$$v(y) := E\int_0^\infty e^{-\lambda t}f(y + B_t)\,dt = E\int_0^\infty e^{-\lambda t}f(y + w_t)\,dt.$$

Upon applying Lemma 3 to $u(x+y)$ in place of $u(x)$, we immediately get that $v=u$, and this proves the theorem if $\lambda>0$.

To prove (3) for $\lambda=0$ it suffices to pass to the limit, which is possible due to the dominated convergence theorem if we know that $E\tau<\infty$. However, for the function $u_0(x)=(x-a)(b-x)$ and the result for $\lambda>0$, we get

$$|a|b=u_0(0)=E\int_0^\tau e^{-\lambda t}(\lambda u(w_t)+1)\,dt\geq E\int_0^\tau e^{-\lambda t}\,dt,$$

$$E\int_0^\tau e^{-\lambda t}\,dt\leq|a|b$$

and it only remains to apply the monotone convergence theorem to get $E\tau\leq|a|b<\infty$. The theorem is proved.

In the following exercises we suggest the reader use Theorem 4.

5. Exercise. (i) Prove that $E\tau=|a|b$.

(ii) By noticing that

$$Eu(w_\tau)=u(b)P(\tau=\tau_b)+u(a)P(\tau<\tau_b)$$

and taking an appropriate function u, show that the probability that the Wiener process hits b before hitting a is $|a|/(|a|+b)$.

6. Exercise. Sometimes one is interested in knowing how much time the Wiener process spends in a subinterval $[c,d]\subset(a,b)$ before exiting from (a,b). Of course, by this time we mean Lebesgue measure of the set $\{t<\tau:w_t\in[c,d]\}$.

(i) Prove that this time equals

$$\gamma:=\int_0^\tau I_{[c,d]}(w_t)\,dt.$$

(ii) Prove that for any Borel nonnegative f we have

$$E\int_0^\tau f(w_t)\,dt=\frac{2}{b-a}\Big(b\int_a^0 f(y)(y-a)\,dy-a\int_0^b f(y)(b-y)\,dy\Big),$$

and find $E\gamma$.

7. Exercise. Define $x_t=w_t+t$, and find the probability that x_t hits b before hitting a.

7. Itô stochastic integral

In Sec. 3 we introduced the stochastic integral of nonrandom functions on $[0, 1]$ against dw_t. It turns out that a slight modification of this procedure allows one to define stochastic integrals of random functions as well. The way we proceed is somewhat different from the traditional one, which will be presented in Sec. 6.1. We decided to give this definition just in case the reader decides to study stochastic integration with respect to arbitrary square integrable martingales.

Let $(w_t, \mathcal{F}_t)$ be a Wiener process in the sense of Definition 5.3, given on a probability space $(\Omega, \mathcal{F}, P)$. To proceed with defining Itô stochastic integral in the framework of Sec. 3 we take

$$X = \Omega \times (0, \infty), \quad \mathfrak{A} = \mathcal{F} \otimes \mathfrak{B}((0, \infty)), \quad \mu = P \times \ell \tag{1}$$

and define Π as the collection of all sets $A \times (s, t]$ where $0 \le s \le t < \infty$ and $A \in \mathcal{F}_s$. Notice that, for $A \times (s, t] \in \Pi$,

$$\mu(A \times (s, t]) = P(A)(t - s) < \infty,$$

so that $\Pi_0 = \Pi$. For $A \times (s, t] \in \Pi$ let

$$\zeta(A \times (s, t]) = (w_t - w_s)I_A.$$

1. Definition. Denote $\mathcal{P} = \sigma(\Pi)$ and call $\mathcal{P}$ the σ-field of *predictable sets*. The functions on $\Omega \times (0, \infty)$ which are $\mathcal{P}$-measurable are called *predictable* (relative to $\mathcal{F}_t$).

By the way, the name "predictable" comes from the observation that the simplest $\mathcal{P}$-measurable functions are indicators of elements of Π which have the form $I_A I_{(s,t]}$ and are left-continuous, thus predictable on the basis of past observations, functions of time.

2. Exercise*. Prove that Π is a π-system, and by relying on Theorem 3.19 conclude that $L_2(\Pi, \mu) = L_2(\mathcal{P}, \mu)$.

3. Theorem. *The function ζ on Π is a random orthogonal measure with reference measure μ, and $E\zeta(\Delta) = 0$ for every $\Delta \in \Pi$.*

Proof. We have to check the conditions of Definition 3.5. Let $\Delta_1 = A_1 \times (t_1, t_2]$, $\Delta_2 = A_2 \times (s_1, s_2] \in \Pi$. Define

$$f_t(\omega) = I_{\Delta_1}(\omega, t) + I_{\Delta_2}(\omega, t)$$

and introduce the points $r_1 \leq ... \leq r_4$ by ordering t_1, t_2, s_1, and s_2. Obviously, for every $t \geq 0$, the functions $I_{\Delta_i}(\omega, t+)$ are $\mathcal{F}_t$-measurable and the same holds for $f_{t+}(\omega)$. Furthermore, for each ω, $f_t(\omega)$ is piecewise constant and left continuous in t. Therefore,

$$f_t(\omega) = \sum_{i=1}^{3} g_i(\omega) I_{(r_i, r_{i+1}]}(t), \tag{2}$$

where the $g_i = f_{r_i+}$ are $\mathcal{F}_{r_i}$-measurable.

It turns out that for every ω

$$\zeta(\Delta_1) + \zeta(\Delta_2) = \sum_{i=1}^{3} g_i(\omega)(w_{r_{i+1}} - w_{r_i}). \tag{3}$$

One can prove (3) in the following way. Fix an ω and define a continuous function A_t, $t \in [r_1, r_4]$, so that A_t is piecewise linear and equals w_{r_i} at all r_i's. Then by integrating through (2) against dA_t, remembering the definition of f_t and the fact that the integral of a sum equals the sum of integrals, we come to (3).

It follows from (3) that

$$E(\zeta(\Delta_1) + \zeta(\Delta_2))^2 = \sum_{i=1}^{3} E g_i^2 (w_{r_{i+1}} - w_{r_i})^2$$

$$+2 \sum_{i<j} E g_i g_j (w_{r_{i+1}} - w_{r_i})(w_{r_{j+1}} - w_{r_j}),$$

where all expectations make sense because $0 \leq f \leq 2$ and $Ew_t^2 = t < \infty$. Remember that $E(w_{r_{j+1}} - w_{r_j}) = 0$ and $E(w_{r_{i+1}} - w_{r_i})^2 = r_{i+1} - r_i$. Also notice that $(w_{r_{i+1}} - w_{r_i})^2$ and g_i^2 are independent by Exercise 5.2 and, for $i < j$, the g_i are $\mathcal{F}_{r_i}$-measurable and $\mathcal{F}_{r_j}$-measurable, owing to $\mathcal{F}_{r_i} \subset \mathcal{F}_{r_j}$, so that $g_i g_j (w_{r_{i+1}} - w_{r_i})$ is $\mathcal{F}_{r_j}$-measurable and hence independent of $w_{r_{j+1}} - w_{r_j}$. Then we see that

$$E(\zeta(\Delta_1) + \zeta(\Delta_2))^2 = \sum_{i=1}^{3} E f_{r_i+}^2 (r_{i+1} - r_i) = E \int_{r_1}^{r_4} f_t^2 \, dt$$

$$= E \int_{r_1}^{r_4} (I_{\Delta_1} + I_{\Delta_2})^2 \, dt = E \int_{r_1}^{r_4} I_{\Delta_1} \, dt + 2E \int_{r_1}^{r_4} I_{\Delta_1 \cap \Delta_2} \, dt + E \int_{r_1}^{r_4} I_{\Delta_2} \, dt$$

$$= \mu(\Delta_1) + 2\mu(\Delta_1 \cap \Delta_2) + \mu(\Delta_2). \tag{4}$$

By plugging in $\Delta_1 = \Delta_2 = \Delta$, we find that $E\zeta^2(\Delta) = \mu(\Delta)$. Then, developing $E(\zeta(\Delta_1) + \zeta(\Delta_2))^2$ and coming back to (4), we get $E\zeta(\Delta_1)\zeta(\Delta_2) = \mu(\Delta_1 \cap \Delta_2)$. Thus by Definition 3.5 the function ζ is a random orthogonal measure with reference measure μ.

The fact that $E\zeta = 0$ follows at once from the independence of $\mathcal{F}_s$ and $w_t - w_s$ for $t \geq s$. The theorem is proved.

Theorem 3 allows us to apply Theorem 3.13. By combining it with Exercise 2 and Remark 3.15 we come to the following result.

4. Theorem. *In notation* (1) *there exists a unique linear isometric operator* $I : L_2(\mathcal{P}, \mu) \to L_2(\mathcal{F}, P)$ *such that, for every* $n = 1, 2, ...$, *constants* c_i, $s_i \leq t_i$, *and* $A_i \in \mathcal{F}_{s_i}$ *given for* $i = 1, ..., n$, *we have*

$$I(\sum_{i=1}^n c_i I_{A_i} I_{(s_i,t_i]}) = \sum_{i=1}^n c_i I_{A_i}(w_{t_i} - w_{s_i}) \quad (a.s.). \tag{5}$$

In addition, $EIf = 0$ *for every* $f \in L_2(\mathcal{P}, \mu)$.

5. Exercise*. Formula (5) admits the following generalization. Prove that for every $n = 1, 2, ...$, constants $s_i \leq t_i$, and $\mathcal{F}_{s_i}$-measurable functions g_i given for $i = 1, ..., n$ and satisfying $Eg_i^2 < \infty$, we have

$$I(\sum_{i=1}^n g_i I_{(s_i,t_i]}) = \sum_{i=1}^n g_i(w_{t_i} - w_{s_i}) \quad (a.s.).$$

6. Definition. We call If, introduced in Theorem 4, *the Itô stochastic integral of* f, and write

$$If =: \int_0^\infty f(\omega, t)\, dw_t.$$

The Itô integral between nonrandom a and b such that $0 \leq a \leq b \leq \infty$ is naturally defined by

$$\int_a^b f(\omega, t)\, dw_t = \int_0^\infty f(\omega, t) I_{(a,b]}(t)\, dw_t.$$

The comments in Sec. 3 before Theorem 3.22 are valid for Itô stochastic integrals as well as for integrals of nonrandom functions against dw_t. It is natural to notice that for nonrandom functions both integrals introduced in this section and in Sec. 3 coincide (a.s.). This follows from formula (3.7), valid for both integrals (and from the possibility of finding appropriate f^n, a possibility which is either known to the reader or will be seen from Remark 8.6).

Generally it is safe to say that the properties of the Itô integral are absolutely different from those of the integral of nonrandom functions. For instance Exercise 3.23 implies that for nonrandom integrands the integral is either zero or its distribution has density. About 1981 M. Safonov constructed an example of random f_t satisfying $1 \leq f_t \leq 2$ and such that the distribution of $\int_0^1 f_t \, dw_t$ is singular with respect to Lebesgue measure.

One may wonder why we took sets like $A \times (s, t]$ and not $A \times [s, t)$ as a starting point for stochastic integration. Actually, for the Itô stochastic integral against the Wiener process this is irrelevant, and the second approach even has some advantages, since then (cf. Exercise 5) almost by definition we would have a very natural formula:

$$\int_0^\infty f(t) \, dw_t = \sum_{i=1}^n f(t_i)(w_{t_{i+1}} - w_{t_i})$$

provided that $f(t)$ is $\mathcal{F}_t$-measurable and $E|f(t)|^2 < \infty$ for every t, and $0 \leq t_1 \leq ... \leq t_{n+1} < \infty$ are nonrandom and such that $f(t) = f(t_i)$ for $t \in [t_i, t_{i+1})$ and $f(t) = 0$ for $t \geq t_{n+1}$. We show that this formula is indeed true in Theorem 8.8.

However, there is a significant difference between the two approaches if one tries to integrate with respect to discontinuous processes. Several unusual things may happen, and we offer the reader the following exercises showing one of them.

7. Exercise. In completely the same way as above one introduces a stochastic integral against $\bar{\pi}_t := \pi_t - t$, where π_t is the Poisson process with parameter 1. Of course, one needs an appropriate filtration of σ-fields $\mathcal{F}_t$ such that π_t is $\mathcal{F}_t$-measurable and $\pi_{t+h} - \pi_t$ is independent of $\mathcal{F}_t$ for all $t, h \geq 0$. On the other hand, one can integrate against $\bar{\pi}_t$ as usual, since this function has bounded variation on each interval $[0, T]$. In connection with this, prove that

$$E(\text{usual}) \int_0^1 \pi_t \, d\bar{\pi}_t \neq 0,$$

so that either π_t is not stochastically integrable or the usual integral is different from the stochastic one. (As follows from Theorem 8.2, the latter is true.)

8. Exercise. In the situation of Exercise 7, prove that for every predictable nonnegative f_t we have

$$E(\text{usual})\int_0^1 f_t\,d\pi_t = E\int_0^1 f_t\,dt.$$

Conclude that π_t is *not* predictable, and is not $\mathcal{P}^\mu$-measurable either.

8. The structure of Itô integrable functions

Dealing with Itô stochastic integrals quite often requires much attention to tiny details, since often what seems true turns out to be absolutely wrong. For instance, we will see below that the function $I_{(0,\infty)}(w_t)I_{(0,1)}(t)$ is Itô integrable and consequently its Itô integral has zero mean. This may look strange due to the following.

Represent the open set $\{t : w_t > 0\}$ as the countable union of disjoint intervals (α_i, β_i). Clearly $w_{\alpha_i} = w_{\beta_i} = 0$, and

$$I_{(0,\infty)}(w_t)I_{(0,1)}(t) = \sum_i I_{(0,1)\cap(\alpha_i,\beta_i)}(t). \tag{1}$$

In addition it looks natural that

$$\int_0^\infty I_{(0,1)\cap(\alpha_i,\beta_i)}(t)\,dw_t = w_{1\wedge\alpha_i} - w_{1\wedge\beta_i}, \tag{2}$$

where the right-hand side is different from zero only if $\alpha_i < 1$, $\beta_i > 1$, and $w_1 > 0$, i.e. if $1 \in (\alpha_i, \beta_i)$. In that case the right-hand side of (2) equals $(w_1)_+$, and since the integral of a sum should be equal to the sum of integrals, formula (1) shows that the Itô integral of $I_{(0,\infty)}(w_t)I_{(0,1)}(t)$ should equal $(w_1)_+$. However, this is impossible since $E(w_1)_+ > 0$.

The contradiction here comes from the fact that the terms in (1) are not Itô integrable and (2) just does not make sense.

One more example of an integral with no sense gives $\int_0^1 w_1\,dw_t$. Again its mean value should be zero, but under every reasonable way of defining this integral it should equal $w_1\int_0^1 dw_t = w_1^2$.

All this leads us to the necessity of investigating the set of Itô integrable functions. Due to Theorem 3.19 and Exercise 3.2 this is equivalent to investigating which functions are $\mathcal{P}^\mu$-measurable.

1. Definition. A function $f_t(\omega)$ given on $\Omega \times (0,\infty)$ is called $\mathcal{F}_t$*-adapted* if it is $\mathcal{F}_t$-measurable for each $t > 0$. By H we denote the set of all real-valued $\mathcal{F}_t$-adapted functions $f_t(\omega)$ which are $\mathcal{F} \otimes \mathfrak{B}(0,\infty)$-measurable and satisfy

$$E \int_0^\infty f_t^2 \, dt < \infty.$$

The following theorem says that all elements of H are Itô integrable. The reader is sent to Sec. 7 for necessary notation.

2. Theorem. *We have* $H \subset L_2(\mathcal{P}, \mu)$.

Proof (Doob). It suffices only to prove that $f \in L_2(\mathcal{P}, \mu)$ for $f \in H$ such that $f_t(\omega) = 0$ for $t \geq T$, where T is a constant. Indeed, by the dominated convergence theorem

$$\int_X |f_t - f_t I_{t \leq n}|^2 \, dP dt = E \int_n^\infty f_t^2 \, dt \to 0$$

as $n \to \infty$, so that, if $f_t I_{t \leq n} \in L_2(\mathcal{P}, \mu)$, then $f_t \in L_2(\mathcal{P}, \mu)$ due to the completeness of $L_2(\mathcal{P}, \mu)$.

Therefore we fix an $f \in H$ and $T < \infty$ and assume that $f_t = 0$ for $t \geq T$. It is convenient to assume that f_t is defined for negative t as well, and $f_t = 0$ for $t \leq 0$. Now we recall that it is known from integration theory that every L_2-function is continuous in L_2. More precisely, if $h \in L_2([0,T])$ and $h(t) = 0$ outside $[0,T]$, then

$$\lim_{a \to 0} \int_{-T}^T |h(t+a) - h(t)|^2 \, dt = 0.$$

This and the inequality

$$\int_{-T}^T |f_{t+a} - f_t|^2 \, dt \leq 2\Big(\int_{-T}^T f_{t+a}^2 \, dt + \int_{-T}^T f_t^2 \, dt \Big) \leq 4 \int_0^T f_t^2 \, dt$$

along with the dominated convergence theorem imply that

$$\lim_{a \to 0} E \int_{-T}^T |f_{t+a} - f_t|^2 \, dt = 0. \tag{3}$$

Now let

$$\rho_n(t) = k2^{-n} \quad \text{for} \quad t \in (k2^{-n}, (k+1)2^{-n}].$$

Changing variables $t + s = u, t = v$ shows that

$$\int_0^1 E\int_0^T |f_{\rho_n(t+s)-s} - f_t|^2\,dtds = \int_0^{T+1}\Big(E\int_{u-1}^{u\wedge T} |f_{\rho_n(u)-u+v} - f_v|^2\,dv\Big)\,du.$$

The last expectation tends to zero owing to (3) uniformly with respect to u, since $0 \le u - \rho_n(u) \le 2^{-n}$. It follows that there is a sequence $n(k) \to \infty$ such that for almost every $s \in [0,1]$

$$\lim_{k\to\infty} E\int_0^T |f_{\rho_{n(k)}(t+s)-s} - f_t|^2\,dt = 0. \tag{4}$$

Fix any s for which (4) holds, and denote $f_t^k = f_{\rho_{n(k)}(t+s)-s}$. Then (4) and the inequality $|a|^2 \le 2|b|^2 + 2|a-b|^2$ show that $|f_t^k|^2$ is μ-integrable at least for all large k.

Furthermore, it turns out that the f_t^k are predictable. Indeed,

$$f_{\rho_n(t+s)-s} = \sum_i f_{i2^{-n}-s} I_{(i2^{-n}-s,(i+1)2^{-n}-s]}(t) = \sum_{i:i2^{-n}-s>0} \cdot \tag{5}$$

In addition, $f_{t_1}I_{(t_1,t_2]}$ is predictable if $0 \le t_1 \le t_2$, since for any Borel B

$$\{(\omega,t) : f_{t_1}(\omega)I_{(t_1,t_2]}(t) \in B\}$$

$$= (\{\omega : f_{t_1}(\omega) \in B\} \times (t_1,t_2]) \cup \{(\omega,t) : I_{(t_1,t_2]}(t) = 0 \in B\} \in \mathcal{P}.$$

Therefore (5) yields the predictability of f_t^k, and the integrability of $|f_t^k|^2$ now implies that $f_t^k \in L_2(\mathcal{P},\mu)$. The latter space is complete, and owing to (4) we have $f_t \in L_2(\mathcal{P},\mu)$. The theorem is proved.

3. Exercise*. By following the above proof, show that left continuous $\mathcal{F}_t$-adapted processes are predictable.

4. Exercise. Go back to Exercise 7.7 and prove that if f_t is left continuous, $\mathcal{F}_t$-adapted, and $E\int_0^1 f_t^2\,dt < \infty$, then the usual integral $\int_0^1 f_t\,d\bar\pi_t$ coincides with the stochastic one (a.s.). In particular, prove that the usual integral $\int_0^1 \pi_{t-}\,d\bar\pi_t$ coincides with the stochastic integral $\int_0^1 \pi_t\,d\bar\pi_t$ (a.s.).

5. Exercise. Prove that if $f \in L_2(\mathcal{P},\mu)$, then there exists $h \in H$ such that $f = h$ μ-a.e. and in this sense $H = L_2(\mathcal{P},\mu)$.

6. Remark. If f_t is independent of ω, (4) implies that for almost any $s \in [0,1]$

$$\lim_{k\to\infty}\int_0^T |f_{\rho_{n(k)}(t+s)-s} - f_t|^2\,dt = 0, \qquad \int_0^T f_t\,dt = \lim_{k\to\infty}\int_0^T f_{\rho_{n(k)}(t+s)-s}\,dt.$$

This means that appropriate Riemann sums converge to the Lebesgue integral of f.

7. Remark. It is seen from the proof of Theorem 2 that, if $f \in H$, then for any integer $n \geq 1$ one can find a partition $0 = t_{n0} < t_{n1} < ... < t_{nk(n)} = n$ such that $\max_i(t_{n,i+1} - t_{ni}) \leq 1/n$ and

$$\lim_{n\to\infty} E\int_0^\infty |f_t - f_t^n|^2\,dt = 0,$$

where $f^n \in H$ are defined by $f_t^n = f_{t_{ni}}$ for $t \in (t_{ni}, t_{n,i+1}]$, $i \leq k(n)-1$, and $f_t^n = 0$ for $t > n$. Furthermore, the f_t^n are predictable, and by Theorem 7.4

$$\int_0^\infty f_t\,dw_t = \operatorname*{l.i.m.}_{n\to\infty} \int_0^\infty f_t^n\,dw_t. \tag{6}$$

One can apply the same construction to vector-valued functions f, and then one sees that the above partitions can be taken the same for any finite number of f's.

Next we prove two properties of the Itô integral. The first one justifies the notation $\int_0^\infty f_t\,dw_t$, and the second one shows a kind of local property of this integral.

8. Theorem. (i) *If $f \in H$, $0 = t_0 < t_1 < ... < t_n < ...$, $f_t = f_{t_i}$ for $t \in [t_i, t_{i+1})$ and $i \geq 0$, then in the mean square sense*

$$\int_0^\infty f_t\,dw_t = \sum_{i=0}^\infty f_{t_i}(w_{t_{i+1}} - w_{t_i}).$$

(ii) *If $g, h \in H$, $A \in \mathcal{F}$, and $h_t(\omega) = g_t(\omega)$ for $t \geq 0$ and $\omega \in A$, then $\int_0^\infty g_t\,dw_t = \int_0^\infty h_t\,dw_t$ on A (a.s.).*

Proof. (i) Define $f_t^i = f_{t_i} I_{(t_i,t_{i+1}]}$ and observe the simple fact that $f = \sum_i f^i$ μ-a.e. Then the linearity and continuity of the Itô integral show that to prove (i) it suffices to prove that

$$\int_0^\infty g I_{(r,s]}(t)\,dw_t = (w_s - w_r)g \tag{7}$$

(a.s.) if g is $\mathcal{F}_r$-measurable, $Eg^2 < \infty$, and $0 \leq r < s < \infty$.

If g is a step function (having the form $\sum_{i=1}^n c_i I_{A_i}$ with constant c_i and $A_i \in \mathcal{F}_r$), then (7) follows from Theorem 7.4. The general case is suggested as Exercise 7.5.

To prove (ii), take common partitions for g and h from Remark 7 and on their basis construct the sequences g_t^n and h_t^n. Then by (i) the left-hand sides of (6) for $f_t^n = g_t^n$ and $f_t^n = h_t^n$ coincide on A (a.s.). Formula (6) then says that the same is true for the integrals of g and h. The theorem is proved.

Much later (see Sec. 6.1) we will come back to Itô stochastic integrals with variable upper limit. We want these integrals to be continuous. For this purpose we need some properties of martingales which we present in the following chapter. The reader can skip it if he/she is only interested in stationary processes.

9. Hints to exercises

2.5 Use Exercise 1.4.14, with $R(x) = x$, and estimate $\int_0^x \sqrt{(-\ln y)/y}\, dy$ through $\sqrt{x(-\ln x)}$ by using l'Hospital's rule.

2.10 The cases $a \leq b$ and $a > b$ are different. At some moment you may like to consult the proof of Theorem 2.3 taking there 2^{2n} in place of n.

2.12 If $P(\xi \leq a, \eta \leq b) = \int_{-\infty}^b f(x)\, dx$ for every b, then $Eg(\eta)I_{\xi \leq a} = \int_{\mathbb{R}} g(x)f(x)\, dx$. The result of these computations is given in Sec. 6.8.

3.4 It suffices to prove that the indicators of sets $(s,t]$ are in $L_p(\Pi, \mu)$.

3.8 Observe that

$$\varphi(s) = E\exp(i\sum_{n=1}^{\infty} f(s+\sigma_n)),$$

and by using the independence of the τ_n and the fact that $EF(\tau_1, \tau_2, ...) = E\Phi(\tau_1)$, where $\Phi(t) = EF(t, \tau_2, ...)$, show that

$$\varphi(s) = \int_0^{\infty} e^{if(s+t)-t}\varphi(s+t)\, dt = e^s \int_s^{\infty} e^{if(t)}(e^{-t}\varphi(t))\, dt.$$

Conclude first that φ is continuous, then that $\varphi(s)e^{-s}$ is differentiable, and solve the above equation. After that, approximate by continuous functions the function which is constant on each interval $(t_j, t_{j+1}]$ and vanishes outside of the union of these intervals.

3.14 Prove that, for every Borel nonnegative f, we have

$$E\sum_{\sigma_n \leq 1} f(\sigma_n) = \int_0^1 f(s)\, ds,$$

and use it to pass to the limit from step functions to arbitrary ones.

3.21 For $b_n > 0$ with $b_n \to 1$, we have $\prod b_n = 0$ if and only if $\sum_n (1 - b_n) = \infty$.

3.23 Use Remark 1.4.10 and (3.6).

5.9 Take any continuous function $u(x)$ defined on $[a, b]$ such that $u < 0$ in (a, b) and $u(a) = u(b) = 0$, and use it to write a formula similar to (5.1).

6.7 Define τ as the first exit time of x_t from (a, b) and, similarly to (6.3), prove that

$$u(0) = E \int_0^{\tau} e^{-\lambda t}(\lambda u(x_t) - u'(x_t) - (1/2)u''(x_t))\, dt + Ee^{-\lambda\tau}u(x_\tau).$$

7.7 Observe that $\int_{(0,t]} \pi_s \, d\pi_s = \pi_t(\pi_t + 1)/2$.

7.8 First take $f_t = I_\Delta$.

8.4 Keep in mind the proof of Theorem 8.2, and redo Exercise 7.5 for π_t in place of w_t.

8.5 Take a sequence of step functions converging to f μ-a.e., and observe that step functions are $\mathcal{F}_t$-adapted.

Chapter 3

Martingales

1. Conditional expectations

The notion of conditional expectation plays a tremendous role in probability theory. In this book it appears in the first place in connection with the theory of martingales, which we will use several times in the future, in particular, to construct a continuous version of the Itô stochastic integral with variable upper limit.

Let $(\Omega, \mathcal{F}, P)$ be a probability space, $\mathcal{G}$ a σ-field and $\mathcal{G} \subset \mathcal{F}$.

1. Definition. Let ξ and η be random variables, and moreover let η be $\mathcal{G}$-measurable. Assume that $E|\xi|, E|\eta| < \infty$ and for every $A \in \mathcal{G}$ we have

$$E\xi I_A = E\eta I_A.$$

Then we call η a *conditional expectation of* ξ *given* $\mathcal{G}$ and write $E\{\xi|\mathcal{G}\} = \eta$. If $\mathcal{G}$ is generated by a random element ζ, one also uses the notation $\eta = E\{\xi|\zeta\}$. Finally, if $\xi = I_A$ with $A \in \mathcal{F}$, then we write $P(A|\mathcal{G}) = E(I_A|\mathcal{G})$.

The notation $E\{\xi|\mathcal{G}\}$ needs a justification.

2. Theorem. *If* η_1 *and* η_2 *are conditional expectations of* ξ *given* $\mathcal{G}$, *then* $\eta_1 = \eta_2$ *(a.s.)*.

Proof. By definition, for any $A \in \mathcal{G}$,

$$E\eta_1 I_A = E\eta_2 I_A, \quad E(\eta_1 - \eta_2)I_A = 0.$$

Since $\eta_1 - \eta_2$ is $\mathcal{G}$-measurable, one can take $A = \{\omega : \eta_1(\omega) - \eta_2(\omega) > 0\}$. Then one gets $E(\eta_1 - \eta_2)_+ = 0$, $(\eta_1 - \eta_2)_+ = 0$, and $\eta_1 \le \eta_2$ (a.s.). Similarly, $\eta_2 \le \eta_1$ (a.s.). The theorem is proved.

The definition of conditional expectation involves expectations. Therefore, if $\eta = E(\xi|\mathcal{G})$, then any $\mathcal{G}$-measurable function coinciding with η almost surely also is a conditional expectation of ξ given $\mathcal{G}$. Theorem 2 says that the converse is also true. To avoid misunderstanding, let us emphasize that if $\eta_1 = E(\xi|\mathcal{G})$ and $\eta_2 = E(\xi|\mathcal{G})$, then we cannot say that $\eta_1(\omega) = \eta_2(\omega)$ for all ω, although this equality does hold for almost every ω.

3. Exercise. Let $\Omega = \bigcup_n A_n$ be a partition of Ω into disjoint sets $A_n \in \mathcal{F}$, $n = 1, 2, \ldots$. Let $\mathcal{G} = \sigma(A_n, n = 1, 2, \ldots)$. Prove that

$$E(\xi|\mathcal{G}) = \frac{1}{P(A_n)} E\xi I_{A_n} \quad \left(\frac{0}{0} := 0\right)$$

almost surely on A_n for any n.

4. Exercise. Let (ξ, η) be an $\mathbb{R}^2$-valued random variable and $p(x, y)$ a nonnegative Borel function on $\mathbb{R}^2$. Remember that p is called a density of (ξ, η) if for any Borel $B \in \mathfrak{B}(\mathbb{R}^2)$ we have

$$P((\xi, \eta) \in B) = \int_B p(x, y)\, dxdy.$$

Denote $\zeta = \int_{\mathbb{R}} p(x, \eta)\, dx$, assume $E|\xi| < \infty$, and prove that (a.s.)

$$E(\xi|\eta) = \frac{1}{\zeta} \int_{\mathbb{R}} xp(x, \eta)\, dx \quad \left(\frac{0}{0} := 0\right).$$

We need some properties of conditional expectations.

5. Theorem. *Let $E|\xi| < \infty$. Then $E(\xi|\mathcal{G})$ exists.*

Proof. On the probability space $(\Omega, \mathcal{G}, P)$ consider the set function $\mu(A) = E\xi_+ I_A$, $A \in \mathcal{G}$. Obviously $\mu \geq 0$ and $\mu(\Omega) = E\xi_+ < \infty$. Furthermore, from measure theory we know that μ is a measure and $\mu(A) = 0$ if $P(A) = 0$. Thus μ is absolutely continuous with respect to P, and by the Radon-Nikodým theorem there is a $\mathcal{G}$-measurable function $\eta_{(+)} \geq 0$ such that

$$\mu(A) = \int_A \eta_{(+)}\, P(d\omega) = E\eta_{(+)} I_A$$

for any $A \in \mathcal{G}$. Similarly, there is a $\mathcal{G}$-measurable $\eta_{(-)}$ such that $E\xi_- I_A = E\eta_{(-)} I_A$ for any $A \in \mathcal{G}$. The random variable $\eta_{(+)} - \eta_{(-)}$ is obviously a conditional expectation of ξ given $\mathcal{G}$. The theorem is proved.

The next theorem characterizes computing conditional expectation as a linear operation.

6. Theorem. *Let $E|\xi| < \infty$. Then*

(i) *for any constant c we have $E(c\xi|\mathcal{G}) = cE(\xi|\mathcal{G})$ (a.s.), in particular, $E(0|\mathcal{G}) = 0$ (a.s.);*

(ii) *we have $EE(\xi|\mathcal{G}) = E\xi$;*

(iii) *if $E|\xi_1|, E|\xi_2| < \infty$, then $E(\xi_1 \pm \xi_2|\mathcal{G}) = E(\xi_1|\mathcal{G}) \pm E(\xi_2|\mathcal{G})$ (a.s.);*

(iv) *if ξ is $\mathcal{G}$-measurable, then $E(\xi|\mathcal{G}) = \xi$ (a.s.);*

(v) *if a σ-field $\mathcal{G}_1 \subset \mathcal{G}$, then*

$$E\big\{E(\xi|\mathcal{G}_1)|\mathcal{G}\big\} = E\big\{E(\xi|\mathcal{G})|\mathcal{G}_1\big\} = E(\xi|\mathcal{G}_1)$$

(a.s.), which can be expressed as the statement that the smallest σ-field prevails.

Proof. Assertions (i) through (iv) are immediate consequences of the definitions and Theorem 2. To prove (v), let $\eta = E(\xi|\mathcal{G})$, $\eta_1 = E(\xi|\mathcal{G}_1)$. Since η_1 is $\mathcal{G}_1$-measurable and $\mathcal{G}_1 \subset \mathcal{G}$, we have that η_1 is $\mathcal{G}$-measurable. Hence $E(\eta_1|\mathcal{G}) = \eta_1$ (a.s.) by (iv); that is,

$$E\big\{E(\xi|\mathcal{G}_1)|\mathcal{G}\big\} = E(\xi|\mathcal{G}_1).$$

Furthermore, if $A \in \mathcal{G}_1$, then $A \in \mathcal{G}$ and $E\eta I_A = E\xi I_A = E\eta_1 I_A$ by definition. The equality of the extreme terms by definition means that $E(\eta|\mathcal{G}_1) = \eta_1$. The theorem is proved.

Next we study the properties of conditional expectations related to inequalities and limits.

7. Theorem. (i) *If $E|\xi_1|, E|\xi_2| < \infty$, and $\xi_2 \geq \xi_1$ (a.s.), then $E(\xi_2|\mathcal{G}) \geq E(\xi_1|\mathcal{G})$ (a.s.).*

(ii) *(The monotone convergence theorem) If $E|\xi_i|, E|\xi| < \infty$, $\xi_{i+1} \geq \xi_i$ (a.s.) for $i = 1, 2, ...$ and $\xi = \lim_{i\to\infty} \xi_i$ (a.s.), then*

$$\lim_{i\to\infty} E(\xi_i|\mathcal{G}) = E(\xi|\mathcal{G}) \quad (a.s.).$$

(iii) *(Fatou's theorem) If $\xi_i \geq 0$, $E\xi_i < \infty$, $i = 1, 2, ...$, and $E \varliminf_{i\to\infty} \xi_i < \infty$, then*

$$E\{\varliminf_{i\to\infty} \xi_i|\mathcal{G}\} \leq \varliminf_{i\to\infty} E\{\xi_i|\mathcal{G}\} \quad (a.s.).$$

(iv) *(The dominated convergence theorem) If $|\xi_i| \leq \eta$, $E\eta < \infty$, and the limit $\lim_{i\to\infty} \xi_i =: \xi$ exists, then*

$$E(\xi|\mathcal{G}) = \lim_{i\to\infty} E(\xi_i|\mathcal{G}) \quad (a.s.).$$

(v) (*Jensen's inequality*) *If* $\phi(t)$ *is finite and convex on* $\mathbb{R}$ *and* $E|\xi| + E|\phi(\xi)| < \infty$, *then* $E(\phi(\xi)|\mathcal{G}) \geq \phi(E(\xi|\mathcal{G}))$ *(a.s.).*

Proof. (i) Let $\eta_i = E(\xi_i|\mathcal{G})$, $A = \{\omega : \eta_2(\omega) - \eta_1(\omega) \leq 0\}$. Then

$$E(\eta_2 - \eta_1)_- = E\eta_1 I_A - E\eta_2 I_A = E\xi_1 I_A - E\xi_2 I_A \leq 0.$$

Hence $E(\eta_2 - \eta_1)_- = 0$ and $\eta_2 \geq \eta_1$ (a.s.).

(ii) Again let $\eta_i = E(\xi_i|\mathcal{G})$. Then the sequence η_i increases (a.s.) and if $\eta := \lim_{i\to\infty} \eta_i$ on the set where the limit exists, then by the monotone convergence theorem

$$E\xi I_A = \lim_{i\to\infty} E\xi_i I_A = \lim_{i\to\infty} E\eta_i I_A = E\eta I_A$$

for every $A \in \mathcal{G}$. Hence by definition $\eta = E(\xi|\mathcal{G})$.

(iii) Observe that if a_n, b_n are two sequences of numbers and the $\lim a_n$ exists and $a_n \leq b_n$, then $\lim a_n \leq \underline{\lim}\, b_n$. Since $\inf(\xi_i, i \geq n)$ increases with n and is less than ξ_n for each n, by the above we have

$$E(\lim_{n\to\infty} \inf(\xi_i, i \geq n)|\mathcal{G}) = \lim_{n\to\infty} E(\inf(\xi_i, i \geq n)|\mathcal{G}) \leq \underline{\lim}_{n\to\infty} E(\xi_n|\mathcal{G}) \quad \text{(a.s.)}.$$

(iv) Owing to (iii),

$$\underline{\lim}_{i\to\infty} E(\xi_i|\mathcal{G}) = \underline{\lim}_{i\to\infty} E(\xi_i + \eta|\mathcal{G}) - E(\eta|\mathcal{G}) \geq E(\xi + \eta|\mathcal{G}) - E(\eta|\mathcal{G}) = E(\xi|\mathcal{G})$$

(a.s.). Upon replacing ξ_i and ξ with $-\xi_i$ and $-\xi$, we also get

$$\overline{\lim_{i\to\infty}} E(\xi_i|\mathcal{G}) \leq E(\xi|\mathcal{G})$$

(a.s.). The combination of these two inequalities proves (iv).

(v) It is well known that there exists a countable set of pairs $(a_i, b_i) \in \mathbb{R}^2$ such that for all t

$$\phi(t) = \sup_i (a_i t + b_i).$$

Hence, for any i, $\phi(t) \geq a_i t + b_i$ and $E(\phi(\xi)|\mathcal{G}) \geq a_i E(\xi|\mathcal{G}) + b_i$ (a.s.). It only remains to take the sup with respect to countably many i's (preserving (a.s.)). The theorem is proved.

8. Corollary. *If* $\xi \geq 0$ *and* $E\xi < \infty$, *then* $E(\xi|\mathcal{G}) \geq 0$ *(a.s.).*

9. Corollary. *If* $p \geq 1$ *and* $E|\xi|^p < \infty$, *then* $|E(\xi|\mathcal{G})|^p \leq E(|\xi|^p|\mathcal{G})$ *(a.s.). In particular,* $|E(\xi|\mathcal{G})| \leq E(|\xi||\mathcal{G})$ *(a.s.).*

10. Corollary. *If $E|\xi| < \infty$ and $E|\xi - \xi_i| \to 0$ as $i \to \infty$, then*

$$E|E(\xi|\mathcal{G}) - E(\xi_i|\mathcal{G})| \leq E|\xi - \xi_i| \to 0.$$

11. Remark. The monotone convergence theorem can be used to define $E(\xi|\mathcal{G})$ as the limit of the increasing sequence $E(\xi \wedge n|\mathcal{G})$ as $n \to \infty$ for any ξ satisfying $E\xi_- < \infty$. With this definition we would not need the condition $E|\phi(\xi)| < \infty$ in Theorem 7 (v), and some other results would hold true under less restrictive assumptions. However, in this book the notation $E(\xi|\mathcal{G})$ is only used for ξ with $E|\xi| < \infty$.

The following theorem shows the relationship between conditional expectations and independence.

12. Theorem. (i) *Let $E|\xi| < \infty$ and assume that ξ and $\mathcal{G}$ are independent (see Definition 2.5.1). Then $E(\xi|\mathcal{G}) = E\xi$ (a.s.). In particular, $E(c|\mathcal{G}) = c$ (a.s.) for any constant c.*

(ii) *Let $E|\xi| < \infty$ and $B \in \mathcal{G}$. Then $E(\xi I_B|\mathcal{G}) = I_B E(\xi|\mathcal{G})$ (a.s.).*

(iii) *Let $E|\xi| < \infty$, let ζ be $\mathcal{G}$-measurable, and let $E|\xi\zeta| < \infty$. Then $E(\xi\zeta|\mathcal{G}) = \zeta E(\xi|\mathcal{G})$ (a.s.).*

Proof. (i) Let $\kappa_n(t) = 2^{-n}[2^n t]$ and $\xi_n = \kappa_n(\xi)$. Take $A \in \mathcal{G}$ and notice that $|\xi - \xi_n| \leq 2^{-n}$. Then our assertion follows from

$$E\xi I_A = \lim_{n\to\infty} E(\xi_n I_A) = \lim_{n\to\infty} \sum_{k=-\infty}^{\infty} \frac{k}{2^n} P(k2^{-n} \leq \xi < (k+1)2^{-n}, A)$$

$$= \lim_{n\to\infty} \sum_{k=-\infty}^{\infty} \frac{k}{2^n} P(k2^{-n} \leq \xi < (k+1)2^{-n}) P(A)$$

$$= P(A) \lim_{n\to\infty} \sum_{k=-\infty}^{\infty} \frac{k}{2^n} P(k2^{-n} \leq \xi < (k+1)2^{-n}) = P(A)E\xi = E(I_A E\xi).$$

(ii) For $\eta = E(\xi|\mathcal{G})$ and any $A \in \mathcal{G}$ we have

$$E(\xi I_B)I_A = E\xi I_{AB} = E\eta I_{AB} = E(\eta I_B)I_A,$$

which yields the result by definition.

(iii) Denote $\zeta_n = \kappa_n(\zeta)$ and observe that $|\zeta_n\xi| \leq |\zeta\xi|$. Therefore, $E\zeta_n\xi$ and $E(\zeta_n\xi|\mathcal{G})$ exist. Also let $B_{nk} = \{\omega : k2^{-n} \leq \zeta < (k+1)2^{-n}\}$. Then

$$I_{B_{nk}} E(\zeta_n\xi|\mathcal{G}) = E(I_{B_{nk}}\zeta_n\xi|\mathcal{G}) = k2^{-n} I_{B_{nk}} E(\xi|\mathcal{G}) = I_{B_{nk}}\zeta_n E(\xi|\mathcal{G})$$

(a.s.). In other words, $E(\zeta_n\xi|\mathcal{G}) = \zeta_n E(\xi|\mathcal{G})$ on B_{nk} (a.s.). Since $\bigcup_k B_{nk} = \Omega$, this equality holds almost surely. By letting $n \to \infty$ and using $|\zeta_n - \zeta| \le 2^{-n}$, we get the result. The theorem is proved.

Sometimes the following generalization of Theorem 12 (iii) is useful.

13. Theorem. *Let $f(x, y)$ be a Borel nonnegative function on $\mathbb{R}^2$, and let ζ be $\mathcal{G}$-measurable and ξ independent of $\mathcal{G}$. Assume $Ef(\xi, \zeta) < \infty$. Denote $\Phi(y) := Ef(\xi, y)$. Then $\Phi(y)$ is a Borel function of y and*

$$E(f(\xi,\zeta)|\mathcal{G}) = \Phi(\zeta) \quad (a.s.). \tag{1}$$

Proof. We just repeat part of the usual proof of Fubini's theorem. Let Λ be the collection of all Borel sets $B \subset \mathbb{R}^2$ such that $EI_B(\xi, y)$ is a Borel function and

$$E(I_B(\xi,\zeta)|\mathcal{G}) = (EI_B(\xi,y))|_{y=\zeta} \quad \text{(a.s.)}. \tag{2}$$

On the basis of the above results it is easy to check that Λ is a λ-system. In addition, Λ contains the π-system Π of all sets $A \times B$ with $A, B \in \mathfrak{B}(\mathbb{R})$, since $I_{A\times B}(\xi, y) = I_B(y)I_A(\xi)$. Therefore, Λ contains the smallest σ-field generated by Π. Since $\sigma(\Pi) = \mathfrak{B}(\mathbb{R}^2)$, $EI_B(\xi, y)$ is a Borel function for all Borel $B \in \mathbb{R}^2$.

Now a standard approximation of nonnegative Borel functions by linear combinations of indicators shows that $\Phi(y)$ is indeed a Borel function and leads from (2) to (1). The theorem is proved.

In some cases one can find conditional expectations by using Exercise 3 and the following result, the second assertion of which is called *the normal correlation theorem.*

14. Theorem. (i) *Let $\mathcal{G}$ be a σ-field, $\mathcal{G} \subset \mathcal{F}$. Denote $H = L_2(\mathcal{F}, P)$, $H_1 = L_2(\mathcal{G}, P)$, and let $\pi_{\mathcal{G}}$ be the orthogonal projection operator of H on H_1. Then, for each random variable ξ with $E\xi^2 < \infty$, we have $E(\xi|\mathcal{G}) = \pi_{\mathcal{G}}\xi$ (a.s.). In particular,*

$$E(\xi - \pi_{\mathcal{G}}\xi)^2 = \inf\{E(\xi - \eta)^2 : \eta \text{ is } \mathcal{G}\text{-measurable}\}.$$

(ii) *Let $(\xi, \xi_1, ..., \xi_n)$ be a Gaussian vector and $\mathcal{G}$ the σ-field generated by $(\xi_1, ..., \xi_n)$. Then $E(\xi|\mathcal{G}) = a + b_1\xi_1 + ... + b_n\xi_n$ (a.s.), where $(a, b_1, ..., b_n)$ is any solution of the system*

$$\begin{aligned} E\xi &= a + b_1 E\xi_1 + ... + b_n E\xi_n, \\ E\xi\xi_i &= aE\xi_i + b_1 E\xi_1\xi_i + ... + b_n E\xi_n\xi_i, \quad i = 1, ..., n. \end{aligned} \tag{3}$$

Furthermore, system (3) always has at least one solution.

Proof. (i) We have that $\pi_{\mathcal{G}}\xi$ is $\mathcal{G}$-measurable, or at least has a $\mathcal{G}$-measurable modification for which we use the same notation. Furthermore, $\xi - \pi_{\mathcal{G}}\xi \perp \eta$ for any $\eta \in H_1$, so that $E(\xi - \pi_{\mathcal{G}}\xi)\eta = 0$. For $\eta = I_A$ with $A \in \mathcal{G}$ this yields $E\xi I_A - EI_A\pi_{\mathcal{G}}\xi = 0$, which by definition means that $E(\xi|\mathcal{G}) = \pi_{\mathcal{G}}\xi$.

(ii) The function $E(\xi-(a+b_1\xi_1+...+b_n\xi_n))^2$ is a nonnegative quadratic function of $(a, b_1, ..., b_n)$.

15. Exercise. Prove that any nonnegative quadratic function attains its minimum at at least one point.

Now take a point $(a, b_1, ..., b_n)$ at which $E(\xi - (a + b_1\xi_1 + ... + b_n\xi_n))^2$ takes its minimum value, and write that all first derivatives with respect to $(a, b_1, ..., b_n)$ vanish at this point. The most convenient way to do this is to express $E(\xi - (a + b_1\xi_1 + ... + b_n\xi_n))^2$ by developing the second power and factoring out all products of constants. Then we will see that system (3) has a solution. Next, notice that for any solution of (3) and $\eta = \xi - (a + b_1\xi_1 + ... + b_n\xi_n)$ we have

$$E\eta\xi_i = 0, \quad E\eta = 0.$$

It follows that in the Gaussian vector $(\eta, \xi_1, ..., \xi_n)$ the first component is uncorrelated with the others. The theory of characteristic functions implies that in that case η is independent of $(\xi_1, ..., \xi_n)$. Since any event $A \in \mathcal{G}$ has the form $\{\omega : (\xi_1, ..., \xi_n) \in \Gamma\}$ with Borel $\Gamma \subset \mathbb{R}^n$, we conclude that η and $\mathcal{G}$ are independent. Hence $E(\eta|\mathcal{G}) = E\eta = 0$ (a.s.), and, adding that ξ_i are $\mathcal{G}$-measurable, we find that

$$0 = E(\eta|\mathcal{G}) = E(\xi|\mathcal{G}) - (a + b_1\xi_1 + ... + b_n\xi_n)$$

(a.s.). The theorem is proved.

16. Remark. Theorem 14 (i) shows another way to introduce the conditional expectations on the basis of Hilbert space theory without using the Radon-Nikodým theorem.

17. Exercise. Let $(\xi, \xi_1, ..., \xi_n)$ be a Gaussian vector with mean zero, and L the set of all linear combinations of ξ_i with constant coefficients. Prove that $E(\xi|\mathcal{G})$ coincides with the orthogonal projection in $L_2(\mathcal{F}, P)$ of ξ on L.

2. Discrete time martingales

A notion close to martingale was used by S.N. Bernstein. According to J. Doob [**Do**] the notion of martingale was introduced in 1939 by J. Ville, who is not very well known in the theory of probability. At the present time the theory of martingales is a very wide and well developed branch of probability theory with many applications in other areas of mathematics. Many mathematicians took part in developing this theory, J. Doob, P. Lévy, P.-A. Meyer, H. Kunita, H. Watanabe, and D. Burkholder should be named in any list of the main contributors to the theory.

Let $(\Omega, \mathcal{F}, P)$ be a complete probability space and let $\mathcal{F}_n$, $n = 1, ..., N$, be a sequence of σ-fields satisfying $\mathcal{F}_1 \subset \mathcal{F}_2 \subset ... \subset \mathcal{F}_N \subset \mathcal{F}$.

1. Definition. A sequence of real-valued random variables ξ_n, $n = 1, ..., N$, such that ξ_n is $\mathcal{F}_n$-measurable and $E|\xi_n| < \infty$ for every n is called

(i) *a martingale* if, for each $1 \leq n \leq m \leq N$,

$$E(\xi_m|\mathcal{F}_n) = \xi_n \quad \text{(a.s.)};$$

(ii) *a submartingale* if, for each $1 \leq n \leq m \leq N$,

$$E(\xi_m|\mathcal{F}_n) \geq \xi_n \quad \text{(a.s.)};$$

(iii) *a supermartingale* if, for each $1 \leq n \leq m \leq N$,

$$E(\xi_m|\mathcal{F}_n) \leq \xi_n \quad \text{(a.s.)}.$$

In those cases in which the σ-field $\mathcal{F}_n$ should be mentioned one says, for instance, that ξ_n is a martingale relative to $\mathcal{F}_n$ or that $(\xi_n, \mathcal{F}_n)$ is a martingale.

Obviously, ξ_n is a supermartingale if and only if $-\xi_n$ is a submartingale, and ξ_n is a martingale if and only if $\pm\xi_n$ are supermartingales. Because of these simple facts we usually state the results only for submartingales or supermartingales, whichever is more convenient. Also trivially, $E\xi_n$ is constant for a martingale, increases with n for submartingales and decreases for supermartingales.

2. Exercise*. By using properties of conditional expectations, prove that:

(i) If $E|\xi| < \infty$, then $\xi_n := E(\xi|\mathcal{F}_n)$ is a martingale.

(ii) If $\eta_1, ..., \eta_N$ are independent, $\mathcal{F}_n = \sigma(\eta_1, ..., \eta_n)$, and $E\eta_n = 0$, then $\xi_n := \eta_1 + ... + \eta_n$ is an $\mathcal{F}_n$-martingale.

(iii) If w_t is a Wiener process and $\mathcal{F}_n = \sigma(w_1, ..., w_n)$, then $(w_n, \mathcal{F}_n)$ and $(\exp(w_n - n/2), \mathcal{F}_n)$ are martingales.

(iv) If ξ_n is a martingale, ϕ is convex and $E|\phi(\xi_n)| < \infty$ for any n, then $\phi(\xi_n)$ is a submartingale. In particular, $|\xi_n|$ is a submartingale.

(v) If ξ_n is a submartingale and ϕ is a convex increasing function satisfying $E|\phi(\xi_n)| < \infty$ for any n, then $\phi(\xi_n)$ is a submartingale. In particular, $(\xi_n)_+$ is a submartingale.

3. Exercise. By Definition 1 and properties of conditional expectations, a sequence of real-valued random variables ξ_n, $n = 1, ..., N$, such that ξ_n is $\mathcal{F}_n$-measurable and $E|\xi_n| < \infty$ for every n, is a martingale if and only if, for each $1 \leq n \leq N$, $\xi_n = E(\xi_N|\mathcal{F}_n)$ (a.s.). This describes all martingales defined for a finite number of times. Prove that a sequence of real-valued random variables $\xi_n \geq 0$, $n = 1, ..., N$, such that ξ_n is $\mathcal{F}_n$-measurable and $E|\xi_n| < \infty$ for every n, is a submartingale if and only if, for each $1 \leq n \leq N$, we have $\xi_n = E(\eta_n|\mathcal{F}_n)$ (a.s.), where η_n is an increasing sequence of nonnegative random variables such that $\eta_N = \xi_N$.

One also has a different characterization of submartingales.

4. Exercise. (i) (Doob's decomposition) Prove that a sequence of real-valued random variables ξ_n, $n = 1, ..., N$, such that ξ_n is $\mathcal{F}_n$-measurable and $E|\xi_n| < \infty$ for every n, is a submartingale if and only if $\xi_n = A_n + m_n$, where m_n is an $\mathcal{F}_n$-martingale and A_n is an increasing sequence such that $A_1 = 0$ and A_n is $\mathcal{F}_{n-1}$-measurable for every $n \geq 2$.

(ii) (Multiplicative decomposition) Prove that a sequence of real-valued random variables $\xi_n \geq 0$, $n = 1, ..., N$, such that ξ_n is $\mathcal{F}_n$-measurable and $E|\xi_n| < \infty$ for every n, is a submartingale if and only if $\xi_n = A_n m_n$, where m_n is a nonnegative $\mathcal{F}_n$-martingale and A_n is an increasing sequence such that $A_1 = 1$ and A_n is $\mathcal{F}_{n-1}$-measurable for every $n \geq 2$.

5. Exercise. As a generalization of Exercise 2 (iii), prove that if $(w_t, \mathcal{F}_t)$ is a Wiener process, $0 \leq t_0 \leq t_1 \leq ... \leq t_N$, and b_n are $\mathcal{F}_{t_n}$-measurable random variables, then

$$\Big(\exp\Big\{\sum_{i=0}^{n-1} b_i(w_{t_{i+1}} - w_{t_i}) - (1/2)\sum_{i=0}^{n-1} b_i^2(t_{i+1} - t_i)\Big\}, \mathcal{F}_{t_n}\Big),$$

$n = 1, ..., N$, is a martingale with expectation 1.

The above definition describes martingales with discrete time parameter. Similarly one introduces martingales defined on any subset of $\mathbb{R}$. A distinguished feature of discrete time martingales is described in the following lemma.

6. Lemma. *$(\xi_n, \mathcal{F}_n)_{n=1}^N$ is a martingale if and only if the ξ_n are $\mathcal{F}_n$-measurable and*

$$E|\xi_n| < \infty \quad \forall n \leq N, \quad E(\xi_{n+1}|\mathcal{F}_n) = \xi_n \quad (a.s.) \quad \forall n \leq N-1.$$

Similar assertions are true for sub- and supermartingales.

Proof. The "only if" part is obvious. To prove the "if" part, notice that, for $m = n$, ξ_n is $\mathcal{F}_n$-measurable and $E(\xi_m|\mathcal{F}_n) = \xi_n$ (a.s.). For $m = n+1$ this equality holds by the assumption. For $m = n+2$, since $\mathcal{F}_n \subset \mathcal{F}_{n+1}$ we have

$$E(\xi_m|\mathcal{F}_n) = E(E(\xi_{n+2}|\mathcal{F}_{n+1})|\mathcal{F}_n) = E(\xi_{n+1}|\mathcal{F}_n) = \xi_n \quad \text{(a.s.)}.$$

In the same way one considers other $m \in \{n, ..., N\}$. The lemma is proved.

7. Definition. Let real-valued random variables ξ_n and σ-fields $\mathcal{F}_n \subset \mathcal{F}$ be defined for $n = 1, 2, ...$ and be such that ξ_n is $\mathcal{F}_n$-measurable and

$$E|\xi_n| < \infty, \quad \mathcal{F}_{n+1} \subset \mathcal{F}_n, \quad E(\xi_n|\mathcal{F}_{n+1}) = \xi_{n+1}$$

(a.s.) for all n. Then we say that $(\xi_n, \mathcal{F}_n)$ is a *reverse martingale*.

An important and somewhat unexpected example of a reverse martingale is given in the following theorem.

8. Theorem. *Let $\eta_1, ..., \eta_N$ be independent identically distributed random variables with $E|\eta_1| < \infty$. Define*

$$\xi_n = (\eta_1 + ... + \eta_n)/n, \quad \mathcal{F}_n = \sigma(\xi_m : m = n, ..., N).$$

Then $(\xi_n, \mathcal{F}_n)$ is a reverse martingale.

Proof. Simple manipulations show that it suffices to prove that

$$E(\eta_i|\xi_n, ..., \xi_N) = \xi_n \quad \text{(a.s.)} \tag{1}$$

for $n = 1, ..., N$ and $i = 1, ..., n$. In turn (1) will be proved if we prove that

$$E(\eta_1|\xi_n, ..., \xi_N) = E(\eta_i|\xi_n, ..., \xi_N) \quad i = 1, ..., n \tag{2}$$

(a.s.). Indeed, then, upon letting $\zeta = E(\eta_1|\xi_n, ..., \xi_N)$, we find that

$$\xi_n = E(\xi_n|\xi_n, ..., \xi_N) = \frac{1}{n} E\big((\eta_1 + ... + \eta_n)|\xi_n, ..., \xi_N\big) = \frac{1}{n} n\zeta$$

(a.s.), which implies (1).

To prove (2), observe that any event $A \in \sigma(\xi_n, ..., \xi_N)$ can be written as $\{\omega : (\xi_n, ..., \xi_N) \in B\}$, where B is a Borel subset of $\mathbb{R}^{N-n+1}$. In addition, the vectors $(\eta_1, \eta_2, ..., \eta_N)$ and $(\eta_2, \eta_1, \eta_3, ..., \eta_N)$ have the same distribution. Therefore, the vectors

$$(\eta_1, \eta_1 + \eta_2 + ... + \eta_n, ..., \eta_1 + \eta_2 + ... + \eta_N)$$

and

$$(\eta_2, \eta_2 + \eta_1 + \eta_3 + ... + \eta_n, ..., \eta_2 + \eta_1 + \eta_3 + ... + \eta_N)$$

have the same distribution. In particular, for $n \geq 2$,

$$E\eta_2 I_{(\xi_n,...,\xi_N)\in B} = E\eta_1 I_{(\xi_n,...,\xi_N)\in B} = E\zeta I_{(\xi_n,...,\xi_N)\in B}.$$

Hence $\zeta = E(\eta_2|\xi_n, ..., \xi_N)$ (a.s.). Similarly one proves (2) for other values of i. The theorem is proved.

3. Properties of martingales

First we adapt the definition of filtration of σ-fields from Sec. 2.7 to the case of sequences.

1. Definition. Let $\mathcal{F}_n$ be σ-fields defined for $n = 0, 1, 2, ...$ and such that $\mathcal{F}_n \subset \mathcal{F}$ and $\mathcal{F}_n \subset \mathcal{F}_{n+1}$. Then we say that we are given an (increasing) *filtration* of σ-fields $\mathcal{F}_n$.

2. Definition. Let τ be a random variable with values in $\{0, 1, ..., \infty\}$. We say that τ is *a stopping time* (relative to $\mathcal{F}_n$) if $\{\omega : \tau(\omega) > n\} \in \mathcal{F}_n$ for all $n = 0, 1, 2, ...$

Observe that we do not assume τ to be finite, on a subset of Ω it may be equal to ∞. The simplest examples of stopping time are given by nonrandom nonnegative integers.

3. Exercise*. Prove that a nonnegative integer-valued random variable τ is a stopping time if and only if $\{\omega : \tau = n\} \in \mathcal{F}_n$ for all $n \geq 0$. Also prove that $\tau \wedge \sigma$, $\tau \vee \sigma$, and $\tau + \sigma$ are stopping times if τ and σ are stopping times.

In applications, quite often the σ-field $\mathcal{F}_n$ is interpreted as the set of all events observable or happening up to moment of time n when we conduct a series of experiments. Assume that we decided to stop our experiments at a random time τ and then, of course, stop observing its future development. Then, for every n, the event $\tau = n$ definitely *either occurs or does not occur* on the interval of time $[0, n]$, which is transformed into the requirement $\{\tau = n\} \in \mathcal{F}_n$. This is the origin of the term "stopping time".

4. Example. Let ξ_n, $n = 0, 1, 2, ...$, be a sequence of $\mathcal{F}_n$-measurable random variables, and let $c \in \mathbb{R}$ be a constant. Define

$$\tau = \inf\{n \geq 0 : \xi_n(\omega) \geq c\} \quad (\inf \emptyset := \infty)$$

as the first time when ξ_n hits $[c, \infty)$ (making the definition $\inf \emptyset := \infty$ natural). It turns out that τ is a stopping time.

Intuitively it is clear, since, for every ω, knowing $\xi_0, ..., \xi_n$ we know whether one of them is higher than c or not, that is, whether $\tau > n$ or not, which shows that $\{\omega : \tau > n\} \in \sigma(\xi_0, ..., \xi_n) \subset \mathcal{F}_n$. To get a rigorous argument, observe that

$$\{\omega : \tau > n\} = \{\omega : \xi_0 < c, ..., \xi_n(\omega) < c\}$$

and this set is in $\mathcal{F}_n$ since, for $i = 0, 1, ..., n$, the ξ_i are $\mathcal{F}_i$-measurable, and because $\mathcal{F}_i \subset \mathcal{F}_n$ they are $\mathcal{F}_n$-measurable as well.

5. Exercise. Let ξ_n be integer valued and c an integer. Assume that τ from Example 4 is finite. Is it true that $\xi_\tau = c$?

If ξ_n, $n = 0, 1, 2, ...$, is a sequence of random variables and τ is an integer-valued variable, then the sequence $\eta_n = \xi_{n\wedge\tau}$ coincides with ξ_n for $n \leq \tau$ and equals ξ_τ after that. Therefore, we say that η_n is the sequence ξ_n *stopped at time* τ.

6. Theorem (Doob). *Let* $(\xi_n, \mathcal{F}_n)$, $n = 0, 1, 2, ...$, *be a submartingale, and let* τ *be an* $\mathcal{F}_n$*-stopping time. Then* $(\xi_{n\wedge\tau}, \mathcal{F}_n)$, $n = 0, 1, 2, ...$, *is a submartingale.*

Proof. Observe that

$$\xi_{n\wedge\tau} = \xi_0 I_{\tau=0} + ... + \xi_n I_{\tau=n} + \xi_n I_{\tau>n}, \quad I_{\tau\leq n}\xi_\tau = \xi_0 I_{\tau=0} + ... + \xi_n I_{\tau=n}.$$

It follows by Exercise 3 that $\xi_{n\wedge\tau}$ and $I_{\tau\leq n}\xi_\tau$ are $\mathcal{F}_n$-measurable. By factoring out $\mathcal{F}_n$-measurable random variables, we find that

$$E(\xi_{(n+1)\wedge\tau}|\mathcal{F}_n) = E(I_{\tau>n}\xi_{n+1}|\mathcal{F}_n) + E(I_{\tau\leq n}\xi_\tau|\mathcal{F}_n) \geq I_{\tau>n}\xi_n + I_{\tau\leq n}\xi_\tau = \xi_{n\wedge\tau}$$

(a.s.). The theorem is proved.

7. Corollary. *If* τ *is a bounded stopping time, then on the set* $\{\omega : \tau \geq n\}$ *we have (a.s.)*

$$E(\xi_\tau|\mathcal{F}_n) \geq \xi_n.$$

Indeed, if $\tau(\omega) \le N$, then $\xi_\tau = \xi_{(N+n)\wedge\tau}$ and

$$E(\xi_\tau|\mathcal{F}_n) = E(\xi_{(N+n)\wedge\tau}|\mathcal{F}_n) \ge \xi_{n\wedge\tau},$$

where the last term equals ξ_n if $\tau \ge n$.

8. Definition. Let τ be a stopping time. Define $\mathcal{F}_\tau$ as the family of all events $A \in \mathcal{F}$ such that

$$A \cap \{\omega : \tau(\omega) \le n\} \in \mathcal{F}_n \quad \forall n = 0, 1, 2, ...$$

9. Exercise*. The notation $\mathcal{F}_\tau$ needs a justification. Prove that, if for an integer n we have $\tau \equiv n$, then $\mathcal{F}_\tau = \mathcal{F}_n$.

Clearly, if $\tau \equiv \infty$, then $\mathcal{F}_\tau = \mathcal{F}$. Also it is not hard to see that $\mathcal{F}_\tau$ is always a σ-field and $\mathcal{F}_\tau \subset \mathcal{F}$. This σ-field is interpreted as the collection of all events which happen during the time interval $[0, \tau]$. The simplest properties of σ-fields $\mathcal{F}_\tau$ are collected in the following lemma.

10. Lemma. *Let τ and σ be stopping times, let ξ, ξ_n, $n = 0, 1, 2, ..., \infty$, be random variables, let ξ_n be $\mathcal{F}_n$-measurable ($\mathcal{F}_\infty := \mathcal{F}$), and let $E|\xi| < \infty$. Then*

(i) $A \in \mathcal{F}_\tau \iff A \cap \{\omega : \tau(\omega) = n\} \in \mathcal{F}_n \quad \forall n = 0, 1, 2, ..., \infty$;

(ii) $\{\omega : \tau \le \sigma\} \in \mathcal{F}_\tau \cap \mathcal{F}_\sigma$ *and, if* $\tau \le \sigma$, *then* $\mathcal{F}_\tau \subset \mathcal{F}_\sigma$;

(iii) τ, ξ_τ, $\xi_n I_{\tau=n}$ *are* $\mathcal{F}_\tau$*-measurable for all* $n = 0, 1, 2, ..., \infty$;

(iv) $E(\xi|\mathcal{F}_\tau) = E(\xi|\mathcal{F}_n)$ (*a.s.*) *on the set* $\{\tau = n\}$ *for any* $n = 0, 1, 2, ..., \infty$.

Proof. We set the proof of (i) as an exercise. To prove (ii) notice that

$$\{\tau \le \sigma\} \cap \{\sigma = n\} = \{\tau \le n\} \cap \{\sigma = n\} \in \mathcal{F}_n,$$

$$\{\tau \le \sigma\} \cap \{\tau = n\} = \{\tau = n, \sigma \ge n\} \in \mathcal{F}_n.$$

Hence $\{\omega : \tau \le \sigma\} \in \mathcal{F}_\tau \cap \mathcal{F}_\sigma$ by (i). In addition, if $\tau \le \sigma$ and $A \in \mathcal{F}_\tau$, then

$$A \cap \{\sigma = n\} = \bigcup_{i=1}^{n} A \cap \{\tau = i\} \cap \{\sigma = n\} \in \mathcal{F}_n$$

because $A \cap \{\tau = i\} \in \mathcal{F}_i \subset \mathcal{F}_n$. Therefore, $A \in \mathcal{F}_\sigma$ for each $A \in \mathcal{F}_\tau$; that is, $\mathcal{F}_\tau \subset \mathcal{F}_\sigma$.

(iii) Since constants are stopping times, (ii) leads to $\{\tau \le n\} \in \mathcal{F}_\tau$, so that τ is $\mathcal{F}_\tau$-measurable. Furthermore, for $A := \{\xi_\tau < c\}$, where c is a constant, we have

$$A \cap \{\tau = n\} = \{\xi_n < c, \tau = n\} \in \mathcal{F}_n$$

for any $n = 0, 1, 2, ..., \infty$. Hence $A \in \mathcal{F}_\tau$ and ξ_τ is $\mathcal{F}_\tau$-measurable. That the same holds for $\xi_n I_{\tau=n}$ follows from $\xi_n I_{\tau=n} = \xi_\tau I_{\tau=n}$.

(iv) Define $\eta = I_{\tau=n} E(\xi|\mathcal{F}_\tau)$ and $\zeta = I_{\tau=n} E(\xi|\mathcal{F}_n)$. Notice that by (i), for any constant c

$$\{\eta < c\} = \{\tau \neq n, 0 < c\} \cup \big(\{\tau = n\} \cap \{E(\xi|\mathcal{F}_\tau) < c\}\big) \in \mathcal{F}_n.$$

Hence η is $\mathcal{F}_n$-measurable. Also ζ is $\mathcal{F}_n$-measurable due to Exercise 3. Furthermore, for any $A \in \mathcal{F}_n$ assertion (iii) (with $\xi_n = I_A$ and $\xi_k = 0$ for $k \neq n$) implies that

$$EI_A\eta = EI_A I_{\tau=n} E(\xi|\mathcal{F}_\tau) = EI_A I_{\tau=n}\xi = EI_A\zeta.$$

Since both η and ζ are $\mathcal{F}_n$-measurable, we conclude that $\eta = E(\eta|\mathcal{F}_n) = \zeta$ (a.s.). The lemma is proved.

11. Theorem (Doob's optional sampling theorem). *Let $(\xi_n, \mathcal{F}_n)$, $n \geq 0$, be a submartingale, and let τ_i, $i = 1, ..., m$, be bounded stopping times satisfying $\tau_1 \leq ... \leq \tau_m$. Then $(\xi_{\tau_i}, \mathcal{F}_{\tau_i})$, $i = 1, ..., m$, is a submartingale.*

Proof. The $\mathcal{F}_{\tau_i}$-measurability of ξ_{τ_i} follows from Lemma 10. This lemma and Corollary 7 imply also that on the set $\{\tau_i = n\}$ we have

$$E(\xi_{\tau_{i+1}}|\mathcal{F}_{\tau_i}) = E(\xi_{\tau_{i+1}}|\mathcal{F}_n) \geq \xi_n = \xi_{\tau_i} \quad \text{(a.s.)}$$

since $\tau_{i+1} \geq n$. Upon noticing that the union of $\{\tau_i = n\}$ is Ω, we get the result. The theorem is proved.

12. Corollary. *If τ and σ are bounded stopping times and $\tau \leq \sigma$, then $\xi_\tau \leq E(\xi_\sigma|\mathcal{F}_\tau)$ and $E\xi_\tau \leq E\xi_\sigma$.*

Surprisingly enough the inequality $E\xi_\tau \leq E\xi_\sigma$ in Corollary 12 can be taken as a definition of submartingale. An advantage of this definition is that it allows one to avoid using the theory of conditional expectations altogether. In connection with this we set the reader the following exercise.

13. Exercise. Let ξ_n be summable $\mathcal{F}_n$-measurable random variables given for $n = 0, 1, ...$ Assume that for any bounded stopping times $\tau \leq \sigma$ we have $E\xi_\tau \leq E\xi_\sigma$, and prove that $(\xi_n, \mathcal{F}_n)$ is a submartingale.

14. Theorem (Doob-Kolmogorov inequality). (i) *If $(\xi_n, \mathcal{F}_n)$, $n = 0, 1, ...,$ is a submartingale, $c > 0$ is a constant, and N is an integer, then*

$$P\{\max_{n \leq N} \xi_n \geq c\} \leq \frac{1}{c} E\xi_N I_{\max_{n \leq N} \xi_n \geq c} \leq \frac{1}{c} E(\xi_N)_+, \tag{1}$$

$$P\{\sup_n \xi_n \geq c\} \leq \frac{1}{c} \sup_n E(\xi_n)_+. \tag{2}$$

(ii) *If* $(\xi_n, \mathcal{F}_n)$, $n = 0, 1, ...,$ *is a supermartingale,* $\xi_n \geq 0$, *and* $c > 0$ *is a constant, then*

$$P\{\sup_n \xi_n \geq c\} \leq \frac{1}{c} E\xi_0.$$

Proof. (i) First we prove (1). Since the second inequality in (1) obviously follows from the first one, we only need to prove the latter. Define

$$\tau = \inf(n \geq 0 : \xi_n \geq c).$$

By applying Corollary 12 with $N \wedge \tau$ and N in place of τ and σ and also using Chebyshev's inequality we find that

$$P\{\max_{n \leq N} \xi_n \geq c\} = P\{\xi_\tau I_{\tau \leq N} \geq c\} \leq \frac{1}{c} E\xi_\tau I_{\tau \leq N} = \frac{1}{c} E\xi_{N\wedge\tau} I_{\tau \leq N\wedge\tau}$$

$$\leq \frac{1}{c} EI_{\tau \leq N\wedge\tau} E(\xi_N | \mathcal{F}_{N\wedge\tau}) = \frac{1}{c} EI_{\tau \leq N\wedge\tau} \xi_N = \frac{1}{c} E\xi_N I_{\max_{n \leq N} \xi_n \geq c}.$$

To prove (2) notice that, for any $\varepsilon > 0$,

$$\{\sup_n \xi_n \geq c\} \subset \bigcup_N \{\max_{n \leq N} \xi_n \geq c - \varepsilon\}$$

and the terms in the union expand as N grows. Hence, for $\varepsilon < c$

$$P\{\sup_n \xi_n \geq c\} \leq \lim_{N\to\infty} P\{\max_{n \leq N} \xi_n \geq c - \varepsilon\}$$

$$\leq \varliminf_{N\to\infty} \frac{1}{c-\varepsilon} E(\xi_N)_+ \leq \frac{1}{c-\varepsilon} \sup_n E(\xi_n)_+.$$

The arbitrariness of ε proves (2).

(ii) Introduce τ as above and fix an integer N. Then, as in the beginning of the proof,

$$P\{\max_{n \leq N} \xi_n \geq c\} \leq \frac{1}{c} E\xi_\tau I_{\tau \leq N} = \frac{1}{c} E\xi_{N\wedge\tau} I_{\tau \leq N} \leq \frac{1}{c} E\xi_{N\wedge\tau} \leq \frac{1}{c} E\xi_0.$$

Now one can let $N \to \infty$ as above. The theorem is proved.

15. Theorem (Doob's inequality). *If $(\xi_n, \mathcal{F}_n)$, $n = 0, 1, ...,$ is a nonnegative submartingale and $p > 1$, then*

$$E\big[\sup_n \xi_n\big]^p \le q^p \sup_n E\xi_n^p, \tag{3}$$

where $q = p/(p-1)$. In particular,

$$E\big[\sup_n \xi_n\big]^2 \le 4 \sup_n E\xi_n^2.$$

Proof. Without losing generality we assume that the right-hand side of (3) is finite. Then for any integer N

$$\big[\sup_{n\le N} \xi_n\big]^p \le \big[\sum_{n\le N} \xi_n\big]^p \le N^p \sum_{n\le N} \xi_n^p, \quad E\big[\sup_{n\le N} \xi_n\big]^p < \infty.$$

Next, by the Doob-Kolmogorov inequality, for $c > 0$,

$$P\{\sup_{n\le N} \xi_n \ge c\} \le \frac{1}{c} E\xi_N I_{\sup_{n\le N} \xi_n \ge c}.$$

We multiply both sides by pc^{p-1}, integrate with respect to $c \in (0, \infty)$, and use

$$P(\eta \ge c) = EI_{\eta \ge c}, \quad \eta^p = p \int_0^\infty c^{p-1} I_{\eta \ge c}\, dc,$$

where η is any nonnegative random variable. We also use Hölder's inequality. Then we find that

$$E\big[\sup_{n\le N} \xi_n\big]^p \le qE\xi_N\big[\sup_{n\le N} \xi_n\big]^{p-1} \le q\big(E\xi_N^p\big)^{1/p}\big(E\big[\sup_{n\le N} \xi_n\big]^p\big)^{1-1/p}.$$

Upon dividing through by the last factor (which is finite by the above) we conclude that

$$E\big[\sup_{n\le N} \xi_n\big]^p \le q^p \sup_n E\xi_n^p.$$

It only remains to use Fatou's theorem and let $N \to \infty$. The theorem is proved.

4. Limit theorems for martingales

Let $(\xi_n, \mathcal{F}_n)$, $n = 0, 1, ..., N$, be a submartingale, and let a and b be fixed numbers such that $a < b$. Define consecutively the following:

$$\tau_1 = \inf(n \geq 0 : \xi_n \leq a) \wedge N, \quad \sigma_1 = \inf(n \geq \tau_1 : \xi_n \geq b) \wedge N,$$

$$\tau_n = \inf(n \geq \sigma_{n-1} : \xi_n \leq a) \wedge N, \quad \sigma_n = \inf(n \geq \tau_n : \xi_n \geq b) \wedge N.$$

Clearly $0 \leq \tau_1 \leq \sigma_1 \leq \tau_2 \leq \sigma_2 \leq ...$ and $\tau_{N+i} = \sigma_{N+i} = N$ for all $i \geq 0$. We have seen before that τ_1 is a stopping time.

1. Exercise*. Prove that all τ_n and σ_n are stopping times.

The points (n, ξ_n) belong to $\mathbb{R}^2$. We join the points (n, ξ_n) and $(n + 1, \xi_{n+1})$ for $n = 0, ..., N - 1$ by straight segments. Then we obtain a piecewise linear function, say l. Let us say that if $\xi_{\tau_m} \leq a$ and $\xi_{\sigma_m} \geq b$, then on $[\tau_m, \sigma_m]$ the function l *upcrosses* (a, b). Denote $\beta(a, b)$ the number of upcrossings of the interval (a, b) by l. It is seen that $\beta(a, b) = m$ if and only if $\xi_{\tau_m} \leq a$, $\xi_{\sigma_m} \geq b$ and either $\xi_{\tau_{m+1}} > a$ or $\xi_{\sigma_{m+1}} < b$.

The following theorem is the basis for obtaining limit theorems for martingales.

2. Theorem (Doob's upcrossing inequality). *If $(\xi_n, \mathcal{F}_n)$, $n = 0, 1, ..., N$, is a submartingale and $a < b$, then*

$$E\beta(a, b) \leq \frac{1}{b - a} E(\xi_N - a)_+.$$

Proof. Notice that $\beta(a, b)$ is also the number of upcrossing of $(0, b - a)$ by the piecewise linear function constructed from $(\xi_n - a)_+$. Furthermore, $\xi_n - a$ and $(\xi_n - a)_+$ are submartingales along with ξ_n. It follows that without loss of generality we may assume that $\xi_n \geq 0$ and $a = 0$. In that case notice that any upcrossing of $(0, b)$ can only occur on an interval of type $[\tau_i, \sigma_i]$ with $\xi_{\sigma_i} - \xi_{\tau_i} \geq b$. Also in any case, $\xi_{\sigma_n} - \xi_{\tau_n} \geq 0$. Hence,

$$b\beta(a, b) \leq (\xi_{\sigma_1} - \xi_{\tau_1}) + (\xi_{\sigma_2} - \xi_{\tau_2}) + ... + (\xi_{\sigma_N} - \xi_{\tau_N}).$$

Furthermore, $\tau_{n+1} \geq \sigma_n$ and $E\xi_{\tau_{n+1}} \geq E\xi_{\sigma_n}$. It follows that

$$\begin{aligned} &bE\beta(a, b) \\ &\leq -E\xi_{\tau_1} + (E\xi_{\sigma_1} - E\xi_{\tau_2}) + (E\xi_{\sigma_2} - E\xi_{\tau_3}) + ... + (E\xi_{\sigma_{N-1}} - E\xi_{\tau_N}) + E\xi_{\sigma_N} \\ &\qquad\qquad \leq E\xi_{\sigma_N} - E\xi_{\tau_1} \leq E\xi_{\sigma_N} = E\xi_N, \end{aligned}$$

thus proving the theorem.

3. Exercise. For $\xi_n \geq 0$ and $a = 0$ it seems that typically $\xi_{\sigma_n} \geq b$ and $\xi_{\tau_{n+1}} = 0$. Then why do we have $E\xi_{\tau_{n+1}} \geq E\xi_{\sigma_n}$?

If we have a submartingale $(\xi_n, \mathcal{F}_n)$ defined for all $n = 0, 1, 2, ...$, then we can construct our piecewise linear function on $(0, \infty)$ and define $\beta_\infty(a, b)$ as the number of upcrossing of (a, b) on $[0, \infty)$ by this function. Obviously $\beta_\infty(a, b)$ is the monotone limit of upcrossing numbers on $[0, N]$. By Fatou's theorem we obtain the following.

4. Corollary. *If $(\xi_n, \mathcal{F}_n)$, $n = 0, 1, 2, ...$, is a submartingale, then*

$$E\beta_\infty(a, b) \leq \frac{1}{b-a} \sup_n E(\xi_n - a)_+ \leq \frac{1}{b-a} (\sup_n E(\xi_n)_+ + |a|).$$

5. Theorem. *Let one of the following conditions hold:*

(i) $(\xi_n, \mathcal{F}_n)$, $n = 0, 1, 2, ...$, is a submartingale and $\sup_n E(\xi_n)_+ < \infty$;

(ii) $(\xi_n, \mathcal{F}_n)$, $n = 0, 1, 2, ...$, is a supermartingale and $\sup_n E(\xi_n)_- < \infty$;

(iii) $(\xi_n, \mathcal{F}_n)$, $n = 0, 1, 2, ...$, is a martingale and $\sup_n E|\xi_n| < \infty$.

Then the limit $\lim_{n\to\infty} \xi_n$ exists with probability one.

Proof. Obviously we only need prove the assertion under condition (i). Define ρ as the set of all rational numbers on $\mathbb{R}$, and notice that almost obviously

$$\{\omega : \overline{\lim_{n\to\infty}} \xi_n(\omega) > \underline{\lim_{n\to\infty}} \xi_n(\omega)\} = \bigcup_{a,b\in\rho, a<b} \{\omega : \beta_\infty(a, b) = \infty\}.$$

Then it only remains to notice that the events on the right have probability zero since

$$E\beta_\infty(a, b) \leq \frac{1}{b-a} (\sup_n E(\xi_n)_+ + |a|) < \infty,$$

so that $\beta_\infty(a, b) < \infty$ (a.s.). The theorem is proved.

6. Corollary. *Any nonnegative supermartingale converges at infinity with probability one.*

7. Corollary (cf. Exercise 2.2). *If $(\xi_n, \mathcal{F}_n)$, $n = 0, 1, 2, ...$, is a martingale and ξ is a random variable such that $|\xi_n| \leq \xi$ for all n and $E\xi < \infty$, then $\xi_n = E(\xi_\infty|\mathcal{F}_n)$ (a.s.), where $\xi_\infty = \lim_{n\to\infty} \xi_n$.*

Indeed, by the dominated convergence theorem for martingales

$$\xi_n = E(\xi_{n+m}|\mathcal{F}_n) = \lim_{m\to\infty} E(\xi_{n+m}|\mathcal{F}_n) = E(\xi_\infty|\mathcal{F}_n).$$

Corollary 7 describes all bounded martingales. The situation with unbounded, even nonnegative, martingales is much more subtle.

8. Exercise. Let $\xi_n = \exp(w_n - n/2)$, where w_t is a Wiener process. By using Corollary 2.4.3, show that $\xi_\infty = 0$, so that $\xi_n > E(\xi_\infty|\mathcal{F}_n)$. Conclude that $E\sup_n \xi_n = \infty$ and, moreover, that for every nonrandom sequence $n(k) \to \infty$, no matter how sparse it is, $E\sup_k \xi_{n(k)} = \infty$.

In the case of reverse martingales one does not need any additional conditions for its limit to exist.

9. Theorem. *Let $(\xi_n, \mathcal{F}_n)$, $n = 0, 1, 2, \ldots$, be a reverse martingale. Then $\lim_{n\to\infty} \xi_n$ exists with probability one.*

Proof. By definition $(\xi_{-n}, \mathcal{F}_{-n})$, $n = \ldots, -2, -1, 0$, is a martingale. Denote by $\beta_N(a,b)$ the number of upcrossing of (a,b) by the piecewise linear function constructed from ξ_{-n} restricted to $[-N, 0]$. By Doob's theorem, $E\beta_N(a,b) \leq (E|\xi_0| + |a|)/(b-a)$. Hence $E \lim_{N\to\infty} \beta_N(a,b) < \infty$, and we get the result as in the proof of Theorem 5.

10. Theorem (Lévy-Doob). *Let ξ be a random variable such that $E|\xi| < \infty$, and let $\mathcal{F}_n$ be σ-fields defined for $n = 0, 1, 2, \ldots$ and satisfying $\mathcal{F}_n \subset \mathcal{F}$.*

(i) *Assume $\mathcal{F}_n \subset \mathcal{F}_{n+1}$ for each n, and denote by $\mathcal{F}_\infty$ the smallest σ-field containing all $\mathcal{F}_n$ ($\mathcal{F}_\infty = \bigvee_n \mathcal{F}_n$). Then*

$$\lim_{n\to\infty} E(\xi|\mathcal{F}_n) = E(\xi|\mathcal{F}_\infty) \quad (a.s.), \tag{1}$$

$$\lim_{n\to\infty} E|E(\xi|\mathcal{F}_n) - E(\xi|\mathcal{F}_\infty)| = 0. \tag{2}$$

(ii) *Assume $\mathcal{F}_n \supset \mathcal{F}_{n+1}$ for all n and denote $\mathcal{F}_\infty = \bigcap_n \mathcal{F}_n$. Then (1) and (2) hold again.*

To prove the theorem we need the following remarkable result.

11. Lemma (Scheffé). *Let ξ, ξ_n, $n = 1, 2, \ldots$, be nonnegative random variables such that $\xi_n \xrightarrow{P} \xi$ and $E\xi_n \to E\xi$ as $n \to \infty$. Then $E|\xi_n - \xi| \to 0$.*

This lemma follows immediately from the dominated convergence theorem and from the relations

$$|\xi - \xi_n| = 2(\xi - \xi_n)_+ - (\xi - \xi_n), \quad (\xi - \xi_n)_+ \le \xi_+, \quad (\xi - \xi_n)_+ \xrightarrow{P} 0.$$

Proof of Theorem 10. (i) Writing $\xi = \xi_+ - \xi_-$ shows that we may concentrate on $\xi \ge 0$. Then Lemma 11 implies that we only need to prove (1).

Denote $\eta = E(\xi|\mathcal{F}_\infty)$ and observe that

$$\eta_n := E(\xi|\mathcal{F}_n) = E(E(\xi|\mathcal{F}_\infty)|\mathcal{F}_n) = E(\eta|\mathcal{F}_n).$$

Therefore it only remains to prove that if η is $\mathcal{F}_\infty$-measurable, $\eta \ge 0$, and $E\eta < \infty$, then $\eta_n := E(\eta|\mathcal{F}_n) \to \eta$ (a.s.).

Obviously $(\eta_n, \mathcal{F}_n)$ is a nonnegative martingale. By Theorem 5 it has a limit at infinity, which we denote η_∞. Since the η_n are $\mathcal{F}_\infty$-measurable, η_∞ is $\mathcal{F}_\infty$-measurable as well. Now for each $k = 0, 1, 2, ...$ and $A \in \mathcal{F}_k$ we have $A \in \mathcal{F}_n$ for all large n, and by Fatou's theorem

$$EI_A\eta_\infty \le \varliminf_{n\to\infty} EI_A\eta_n = \varliminf_{n\to\infty} EI_A E(\eta|\mathcal{F}_n) = EI_A\eta. \tag{3}$$

Hence $EI_A(\eta - \eta_\infty)$ is a nonnegative measure defined on the algebra $\bigcup_n \mathcal{F}_n$. This measure uniquely extends to $\mathcal{F}_\infty$ and yields a nonnegative measure on $\mathcal{F}_\infty$. Since $EI_A(\eta - \eta_\infty)$ considered on $\mathcal{F}_\infty$ is obviously one of the extensions, we have $EI_A(\eta - \eta_\infty) \ge 0$ for all $A \in \mathcal{F}_\infty$. Upon taking $A = \{\omega : \eta - \eta_\infty \le 0\}$, we see that $E(\eta - \eta_\infty)_- = 0$, that is, $\eta \ge \eta_\infty$ (a.s.).

Furthermore, if η is bounded, then the inequality in (3) becomes an equality, implying $\eta = \eta_\infty$ (a.s.). Thus, in general $\eta \ge \eta_\infty$ (a.s.) and, if η is bounded, then $\eta = \eta_\infty$ (a.s.). It only remains to notice that, for any constant $a \ge 0$,

$$\eta \ge \eta_\infty = \lim_{n\to\infty} E(\eta|\mathcal{F}_n) \ge \lim_{n\to\infty} E(\eta \wedge a|\mathcal{F}_n) = \eta \wedge a \quad \text{(a.s.)},$$

that is, $\eta \ge \eta_\infty \ge \eta \wedge a$ (a.s.), and let $a \to \infty$. This proves (i).

(ii) As in (i) we may and will assume that $\xi \ge 0$. Denote $\xi_n = E(\xi|\mathcal{F}_n)$. Then $(\xi_n, \mathcal{F}_n)$ is a reverse martingale, and $\lim_{n\to\infty} \xi_n$ exists with probability one. We define ξ_∞ to be this limit where it exists, and 0 otherwise. Obviously ξ_∞ is $\mathcal{F}_n$-measurable for any n, and therefore $\mathcal{F}_\infty$-measurable. Let $\hat{\xi} := E(\xi|\mathcal{F}_\infty)$ and for any $A \in \mathcal{F}_\infty$ write

$$EI_A\xi \le \varliminf_{n\to\infty} EI_A\xi_n = EI_A\xi = EI_A\hat{\xi}.$$

It follows that $\xi_\infty \leq \hat{\xi}$ (a.s.). Again, if ξ is bounded, then $\xi_\infty = \hat{\xi}$ (a.s.). Next,

$$\hat{\xi} \geq \xi_\infty = \lim_{n\to\infty} E(\xi|\mathcal{F}_n) \geq \lim_{n\to\infty} E(\xi \wedge a|\mathcal{F}_n) = E(\xi \wedge a|\mathcal{F}_\infty)$$

(a.s.). By letting $a \to \infty$ and using the monotone convergence theorem we conclude that

$$\hat{\xi} \geq \xi_\infty \geq \lim_{a\to\infty} E(\xi \wedge a|\mathcal{F}_\infty) = E(\xi|\mathcal{F}_\infty) = \hat{\xi}$$

(a.s.). The theorem is proved.

From the Lévy-Doob theorem one gets one more proof of the strong law of large numbers.

12. Theorem (Kolmogorov). *Let $\eta_1, \eta_2, ...$ be independent identically distributed random variables with $E|\eta_1| < \infty$. Denote $m = E\eta_1$. Then*

$$\lim_{n\to\infty} \frac{1}{n}(\eta_1 + ... + \eta_n) = m \quad (a.s.), \qquad \lim_{n\to\infty} E\Big|\frac{1}{n}(\eta_1 + ... + \eta_n) - m\Big| = 0.$$

Proof. Without losing generality we assume that $m = 0$. Define

$$\xi_n = (\eta_1 + ... + \eta_n)/n, \quad \mathcal{F}_n^N = \sigma(\xi_n, ..., \xi_N), \quad \mathcal{F}_n = \bigvee_{N\geq n} \mathcal{F}_n^N.$$

We know that $(\xi_n, \mathcal{F}_n^N)$, $n = 1, 2, ..., N$, is a reverse martingale (Theorem 2.8). In particular, $\xi_n = E(\xi_1|\mathcal{F}_n^N)$ (a.s.), whence by Lévy's theorem $\xi_n = E(\xi_1|\mathcal{F}_n)$ (a.s.). Again by Lévy's theorem, $\zeta := \lim_{n\to\infty} \xi_n$ exists almost surely and in $L_1(\mathcal{F}, P)$. It only remains to prove that $\zeta = 0$ (a.s.).

Since $E|\eta_1| < \infty$ and $E\eta_1 = 0$, the function $\phi(t) = Ee^{it\eta_1}$ is continuously differentiable and $\phi'(0) = 0$. In particular, $\phi(t) = 1 + o(t)$ as $t \to 0$ and

$$Ee^{it\zeta} = \lim_{n\to\infty} (\phi(t/n))^n = \lim_{n\to\infty} (1 + o(t/n))^n = 1$$

for any t. This implies $\zeta = 0$ (a.s.). The theorem is proved.

One application of martingales outside of probability theory is related to differentiating.

13. Exercise. Prove the following version of Lebesgue's differentiation theorem. Let $f(t)$ be a finite monotone function on $[0, 1]$. For $x \in [0, 1]$ and integer $n \geq 0$ write $x = k2^{-n} + \varepsilon$, where k is an integer and $0 \leq \varepsilon < 2^{-n}$, and define $a_n(x) = k2^{-n}$ and $b_n(x) = (k+1)2^{-n}$. Prove that

$$\lim_{n\to\infty} \frac{f(b_n(x)) - f(a_n(x))}{b_n(x) - a_n(x)}$$

exists for almost every $x \in [0,1]$.

The following exercise bears on a version of Lebesgue's differentiation theorem for measures.

14. Exercise. Let $(\Omega, \mathcal{F})$ be a measurable space and $(\mathcal{F}_n)$ an increasing filtration of σ-fields $\mathcal{F}_n \subset \mathcal{F}$. Assume $\mathcal{F} = \bigvee_n \mathcal{F}_n$. Let μ and ν be two probability measures on $(\Omega, \mathcal{F})$. Denote by μ_n and ν_n the restrictions of μ and ν respectively on $(\Omega, \mathcal{F}_n)$, and show that for any n and nonnegative $\mathcal{F}_n$-measurable function f we have

$$\int_\Omega f \, \nu(d\omega) = \int_\Omega f \, \nu_n(d\omega). \tag{4}$$

Next, assume that ν_n is absolutely continuous with respect to μ_n and let $\rho_n(\omega)$ be the Radon-Nikodým derivative $\nu_n(d\omega)/\mu_n(d\omega)$. Prove that:

(i) $\lim_{n\to\infty} \rho_n$ exists μ-almost everywhere and, if we denote

$$\rho(\omega) = \begin{cases} \overline{\lim_{n\to\infty}} \, \rho_n(\omega) & \text{if} \quad \rho^* := \sup_n \rho_n < \infty, \\ \infty & \text{otherwise,} \end{cases}$$

then

(ii) if ν is absolutely continuous with respect to μ, then $\rho = \nu(d\omega)/\mu(d\omega)$ and $\rho_n \to \rho$ in $L_1(\mathcal{F}, \mu)$, while

(iii) in the general case ν admits the following decomposition into the sum of absolutely continuous and singular parts: $\nu = \nu_a + \nu_s$, where

$$\nu_a(A) = \int_\Omega I_A \rho \, \mu(d\omega), \quad \nu_s(A) = \nu(A \cap \{\rho = \infty\}).$$

5. Hints to exercises

1.4 Notice that, for any Borel $f(y)$ with $E|f(\eta)| < \infty$, we have

$$Ef(\eta) = \int_{\mathbb{R}^2} f(y) p(x,y) \, dx dy.$$

2.2 In (iii) notice that $w_m = w_n + (w_m - w_n)$, where $w_m - w_n$ is independent of $\mathcal{F}_n$.

2.3 In the proof of necessity start by writing $(0 \cdot 0^{-1} := 0)$

$$\xi_n = \xi_n \big(E(\xi_{n+1}|\mathcal{F}_n)\big)^{-1} E(\xi_{n+1}|\mathcal{F}_n) = \zeta_n E(\xi_{n+1}|\mathcal{F}_n) = E(\zeta_n \xi_{n+1}|\mathcal{F}_n),$$

where $\zeta_n := \xi_n \big(E(\xi_{n+1}|\mathcal{F}_n)\big)^{-1} \le 1$, then iterate.

2.4 (ii) If the decomposition exists, then $E(\xi_n|\mathcal{F}_{n-1}) = A_n m_{n-1}$.

2.5 Use Theorem 1.13.

3.3 Consider the event $\{\tau \leq n\}$.

3.5 In some cases the answer is "no".

3.13 For $A \in \mathcal{F}_n$ define $\tau = n$ on A and $\tau = n+1$ on A^c.

4.13 On the probability space $([0,1], \mathfrak{B}([0,1]), \ell)$ take the filtration of σ-fields $\mathcal{F}_n$ each of which is defined as the σ-field generated by the sets

$$[k2^{-n}, (k+1)2^{-n}), \quad k = 0, 1, ..., 2^n - 1.$$

Then check that $\{f(b_n(x)) - f(a_n(x))\}2^n$ is a martingale relative to $\mathcal{F}_n$.

4.14 (i) By using (4.4) prove that ρ_n is an $\mathcal{F}_n$-martingale on $(\Omega, \mathcal{F}, \mu)$. (iii) For each $a > 0$ define $\tau_a = \inf\{n \geq 0 : \rho_n > a\}$ and show that for every n, $A \in \mathcal{F}_n$, and $m \geq n$

$$\nu(A) = \nu_n(A) = \int_A I_{\tau_a > m}(\rho_m \wedge a)\, \mu(d\omega) + \nu(A \cap \{\tau_a \leq m\}).$$

By letting $m \to \infty$, derive that

$$\nu(A) = \int_A I_{\rho^* \leq a} \rho\, \mu(d\omega) + \nu(A \cap \{\rho^* > a\}).$$

Next let $a \to \infty$ and extend the formula from $A \in \mathcal{F}_n$ to all of $\mathcal{F}$. (ii) Use (iii), remember that ν is a probability measure, and use Scheffé's lemma.

Chapter 4

Stationary Processes

1. Simplest properties of second-order stationary processes

1. Definition. Let $T \in [-\infty, \infty)$. A complex-valued random process ξ_t defined on (T, ∞) is called *second-order stationary* if $E|\xi_t|^2 < \infty$, $E\xi_t$ is constant, and the function $E\xi_s\bar{\xi}_t$ depends only on the difference $s - t$ for $t, s > T$.

The function $R(s-t) = E\xi_s\bar{\xi}_t$ is called *the correlation function* of ξ_t. We will always assume that $E\xi_t \equiv 0$ and that $R(t)$ is continuous in t.

2. Exercise*. Prove that R is continuous if and only if the function ξ_t is continuous in t in the mean-square sense, that is, as a function from (T, ∞) to $L_2(\mathcal{F}, P)$.

Notice some simple properties of R. Obviously, $R(0) = E\xi_t\bar{\xi}_t = E|\xi_t|^2$ is a real number. Also $R(t) = E\xi_t\bar{\xi}_0$ if $0 \in (T, \infty)$, and generally

$$R(t) = E\xi_{r+t}\bar{\xi}_r, \quad R(-t) = E\overline{\bar{\xi}_{s-t}\xi_s} = \overline{E\xi_s\bar{\xi}_{s-t}} = \bar{R}(t) \tag{1}$$

provided $r, r+t, s, s-t \in (T, \infty)$. The most important property of R is that it is positive definite.

3. Definition. A complex-valued function $r(t)$ given on $(-\infty, \infty)$ is called *positive definite* if for every integer $n \geq 1$, $t_1, ..., t_n \in \mathbb{R}$, and complex $z_1, ..., z_n$ we have

$$\sum_{j,k=1}^{n} r(t_j - t_k) z_j \bar{z}_k \geq 0 \tag{2}$$

(in particular, it is assumed that the sum in (2) is a real number).

That R is positive definite, one proves in the following way: take s large enough and write

$$\sum_{j,k=1}^{n} R(t_j - t_k) z_j \bar{z}_k = E \sum_{j,k=1}^{n} z_j \bar{z}_k \xi_{s+t_j} \bar{\xi}_{s+t_k} = E \Big| \sum_{j=1}^{n} z_j \xi_{s+t_j} \Big|^2 \geq 0.$$

Below we prove the Bochner-Khinchin theorem on the general form of positive definite functions. We need the following.

4. Lemma. *Let $r(t)$ be a continuous positive definite function. Then*

(i) $r(0) \geq 0$,

(ii) $\bar{r}(t) = r(-t)$, $|r(t)| \leq r(0)$ and, in particular, $r(t)$ is a bounded function,

(iii) *if $\int_{-\infty}^{\infty} |r(t)|\, dt < \infty$, then*

$$\int_{-\infty}^{\infty} r(t)\, dt \geq 0,$$

(iv) *for every $x \in \mathbb{R}$, the function $e^{itx} r(t)$, as a function of t, is positive definite.*

Proof. Assertion (i) follows from (2) with $n = 1$, $z = 1$. Assertion (iv) also trivially follows from (2) if one replaces z_k with $z_k e^{it_k x}$.

To prove (ii), take $n = 2$, $t_1 = t, t_2 = 0$, $z_1 = z$, $z_2 = \lambda$, where λ is a real number. Then (2) becomes

$$r(0)(|z|^2 + \lambda^2) + \lambda r(t) z + \lambda r(-t) \bar{z} \geq 0. \tag{3}$$

It follows immediately that $r(t)z + r(-t)\bar{z}$ is real for any complex z. Furthermore, since $r(-t)\bar{z} + \bar{r}(-t)z = 2\mathrm{Re}\, r(-t)\bar{z}$ is real, the number $(r(t) - \bar{r}(-t))z$ is real for any complex z, which is only possible when $r(t) - \bar{r}(-t) = 0$.

Next, from (3) with $z = \bar{r}(t)$ we get

$$r(0)|r(t)|^2 + r(0)\lambda^2 + 2\lambda |r(t)|^2 \geq 0$$

for all real λ. It follows that $|r(t)|^4 - r^2(0)|r(t)|^2 \le 0$. This proves assertion (ii).

Turning to assertion (iii), remember that r is continuous and its integral is the limit of appropriate sums. Viewing dt and ds as z_j and $\bar{z}_k$, respectively, from (2), we get

$$\int_{-N}^{N}\int_{-N}^{N} r(t-s)\,dtds \sim \sum_{i,j} r(t_i - t_j)\Delta t_i \Delta t_j \ge 0,$$

$$0 \le \frac{1}{N}\int_{-N}^{N}\int_{-N}^{N} r(t-s)\,dtds = \int_{-\infty}^{\infty} r(t)(2 - \tfrac{|t|}{N})I_{|t|\le 2N}\,dt,$$

where the equality follows after the change of variables $t-s=t', t+s=s'$. By the Lebesgue dominated convergence theorem the last integral converges to $2\int_{-\infty}^{\infty} r(t)\,dt$. This proves assertion (iii) and finishes the proof of the lemma.

5. Theorem (Bochner-Khinchin). *Let $r(t)$ be a continuous positive definite function. Then there exists a unique nonnegative measure F on $\mathbb{R}$ such that $F(\mathbb{R}) = r(0)$ and*

$$r(t) = \int_{\mathbb{R}} e^{itx} F(dx) \quad \forall t \in \mathbb{R}. \tag{4}$$

Proof. The uniqueness follows at once from the theory of characteristic functions. In the proof of existence, without loss of generality, one may assume that $r(0) \ne 0$ and even that $r(0) = 1$.

Assuming that $r(0) = 1$, we first prove (4) in the particular case in which

$$\int_{\mathbb{R}} |r(t)|\,dt < \infty. \tag{5}$$

Then by Lemma 4 (ii) we have

$$\int_{\mathbb{R}} |r(t)|^2\,dt < \infty.$$

Next, define f as the Fourier transform of r:

$$f(x) = \frac{1}{2\pi}\int_{\mathbb{R}} e^{-itx} r(t)\,dt.$$

By Lemma 4 (iii), (iv) we have $f(x) \ge 0$. From the theory of the Fourier transform we obtain that $f \in L_2(\mathbb{R})$ and

$$r(t) = \int_{\mathbb{R}} e^{itx} f(x)\,dx \tag{6}$$

for almost all t, where the last integral is understood in the sense of L_2 (as the limit in L_2 of $\int_{|x|\leq n} e^{itx} f(x)\,dx$). To finish the proof of the theorem in our particular case, we prove that f is integrable, so that the integral in (6) exists in the usual sense and is a continuous function of t, which along with the continuity of r implies that (6) holds for all t rather than only almost everywhere.

By Parseval's identity, for $s > 0$,

$$\int_{\mathbb{R}} e^{-sx^2/2} f(x)\,dx = \frac{1}{\sqrt{2\pi s}} \int_{\mathbb{R}} e^{-t^2/(2s)} r(t)\,dt$$

(knowing the characteristic function of the normal law, we know that the function

$$\sqrt{s/(2\pi)}e^{-sx^2/2}$$

is the Fourier transform of $e^{-t^2/(2s)}$). The last integral is rewritten as $Er(\sqrt{s}\,\xi)$, where $\xi \sim N(0,1)$, and it is seen that, owing to boundedness and continuity of r, this integral converges to $r(0)$ as $s \downarrow 0$. Now the monotone convergence theorem ($f \geq 0$) shows that

$$\int_{\mathbb{R}} f(x)\,dx = r(0) < \infty.$$

This proves the theorem under condition (5).

In the general case, for $\varepsilon > 0$, define

$$r_\varepsilon(t) := r(t)e^{-\varepsilon^2 t^2/2} = Er(t)e^{it\varepsilon\xi}.$$

The second equality and Lemma 4 (iv) show that r_ε is positive definite. Since $r_\varepsilon(0) = 1$ and $\int_{\mathbb{R}} |r_\varepsilon|\,dt < \infty$, there exists a distribution for which r_ε is the characteristic function. Now remember that in probability theory one proves that if a sequence of characteristic functions converges to a function which is continuous at zero, then this function is also the characteristic function of a distribution. Since obviously $r_\varepsilon \to r$ as $\varepsilon \downarrow 0$, the above-mentioned fact brings the proof of our theorem to an end.

6. Definition. The measure F, corresponding to R, is called *the spectral measure* of R or of the corresponding second-order stationary process. If F is absolutely continuous, its density is called *a spectral density* of R.

From the first part of the proof of Theorem 5 we get the following.

7. Corollary. *If* $\int_{\mathbb{R}} |R(t)|\,dt < \infty$, *then* R *admits a bounded continuous spectral density.*

From the uniqueness of representation and from (1) one easily obtains the following.

8. Corollary. *If* R *is real valued* ($\bar{R} = R$) *and the spectral density* f *exists, then* R *is even and* f *is even* ($f(x) = f(-x)$ (*a.e.*)). *Conversely, if* f *is even, then* R *is real valued and even.*

Yet another description of positive definite functions is given in the following theorem.

9. Theorem. *A function* $r(t)$ *is continuous and positive definite if and only if it is the correlation function of a second-order stationary process.*

Proof. The sufficiency has been proved above. While proving the necessity, without loss of generality, we may and will assume that $r(0) = 1$. By the Bochner-Khinchin theorem the spectral distribution F exists. By Theorem 1.1.12 there exists a random variable ξ with distribution F and characteristic function r. Finally, take a random variable φ uniformly distributed on $[-\pi, \pi]$ and independent of ξ, and define

$$\xi_t = e^{i(\xi t + \varphi)}, \quad t \in \mathbb{R}.$$

Then

$$E\xi_t = r(t) E e^{i\varphi} = 0, \quad E\xi_s \bar{\xi}_t = E e^{i\xi(s-t)} = r(s-t),$$

which proves the theorem.

10. Remark. We have two representations for correlation functions of second-order stationary processes:

$$R(s-t) = E\xi_s \bar{\xi}_t \quad \text{and} \quad R(s-t) = \int_{\mathbb{R}} e^{isx} \overline{e^{itx}}\, F(dx).$$

Hence, in some sense, the random variable ξ_t given on Ω corresponds to the function e^{itx} on $\mathbb{R}$. We will see in the future that this correspondence turns out to be very deep.

11. Exercise. In a natural way one gives the definition of a second-order stationary sequence ξ_n given only for integers $n \in (T, \infty)$. For a second-order stationary sequence ξ_n its correlation function $R(n)$ is defined on integers $n \in \mathbb{R}$. Prove that, for each such $R(n)$, there exists a nonnegative measure F on $[-\pi, \pi]$ such that $R(n) = \int_{-\pi}^{\pi} e^{inx} F(dx)$ for all integers $n \in \mathbb{R}$.

Various representation formulas play an important role in the theory of second-order stationary processes. We are going to prove several of them, starting with the following.

12. Theorem (Kotel'nikov-Shannon). *Let the spectral measure F of a second-order stationary process ξ_t, given on $\mathbb{R}$, be concentrated on $(-\pi,\pi)$, so that $F(-\pi,\pi)=F(\mathbb{R})$. Then for every t*

$$\xi_t=\sum_{n=-\infty}^{\infty}\frac{\sin\pi(t-n)}{\pi(t-n)}\xi_n\qquad\Big(\frac{\sin 0}{0}:=1\Big),$$

which is understood as

$$\xi_t=\operatorname*{l.i.m.}_{m\to\infty}\sum_{n=-m}^{m}\frac{\sin\pi(t-n)}{\pi(t-n)}\xi_n.$$

Proof. We have to prove that

$$\lim_{m\to\infty}E\Big|\xi_t-\sum_{n=-m}^{m}\frac{\sin\pi(t-n)}{\pi(t-n)}\xi_n\Big|^2=0. \tag{7}$$

This equality can be expressed in terms of the correlation function alone. It follows that we need only prove (7) for some second-order stationary process with the same correlation function R. We choose the process from Theorem 9. Then we see that the expression in (7) under the limit sign equals

$$E\Big|e^{i\eta t}-\sum_{n=-m}^{m}\frac{\sin\pi(t-n)}{\pi(t-n)}e^{i\eta n}\Big|^2=0. \tag{8}$$

Since the function e^{itx} is continuously differentiable in x, its partial Fourier sums are uniformly bounded and converge to e^{itx} on $(-\pi,\pi)$. The random variable η takes values in $(-\pi,\pi)$ by assumption, and the sum in (8) is a partial Fourier sum of e^{itx} evaluated at $x=\eta$. Now the assertion of the theorem follows from the Lebesgue dominated convergence theorem.

13. Exercise. Let φ be a uniformly distributed random variable on $(-\pi,\pi)$ and $\xi_t:=e^{i(\pi t+\varphi)}$, so that the corresponding spectral measure is concentrated at π. Prove that (for all ω)

$$\sum_{n=-\infty}^{\infty}\frac{\sin\pi(t-n)}{\pi(t-n)}\xi_n=e^{i\varphi}\cos\pi t\neq\xi_t.$$

14. Exercise*. Remember the way the one-dimensional Riemann integral of continuous functions is defined. It turns out that this definition is easily extendible to continuous functions with values in Banach spaces. We mean the following definition.

Let $f(t)$ be a continuous function defined on a finite interval $[0,1]$ with values in a Banach space H. Then

$$\int_0^1 f(t)\,dt := \lim_{n\to\infty} \sum_{i=0}^{2^n-1} f(i2^{-n})2^{-n},$$

where the limit is understood in the sense of convergence in H. Similarly one defines the integrals over finite intervals $[a,b]$. Prove that the limit indeed exists.

15. Exercise. The second-order stationary processes that we concentrate on are assumed to be continuous as $L_2(\mathcal{F},P)$-valued functions. Therefore, according to Exercise 14 for finite a and b, the integral $\int_a^b \xi_t\,dt$ is well-defined as the integral of an $L_2(\mathcal{F},P)$-valued continuous function ξ_t. We say that this is the *mean-square integral*.

By using the same method as in the proof of Theorem 9, prove that if ξ_t is a second-order stationary process defined for all t, then

$$\underset{T\to\infty}{\text{l.i.m.}}\ \frac{1}{2T}\int_{-T}^{T} \xi_t\,dt$$

always exists. Also prove that this limit equals zero if and only if $F\{0\}=0$. Finally prove that $F\{0\}=0$ if $R(t)\to 0$ as $t\to\infty$.

2. Spectral decomposition of trajectories

H. Cramér discovered a representation of trajectories of second-order stationary processes as "sums" of harmonics with random amplitudes. To prove his result we need the following.

1. Lemma. *Let F be a finite measure on $\mathfrak{B}(\mathbb{R})$. Then the set of all functions*

$$f(x) = \sum_{j=1}^{n} c_j e^{it_j x}, \tag{1}$$

where c_j, t_j, and n are arbitrary, is everywhere dense in $L_2(\mathfrak{B}(\mathbb{R}),F)$.

Proof. If the assertion of the lemma is false, then there exists a nonzero element $g \in L_2(\mathfrak{B}(\mathbb{R}), F)$ such that

$$\int_{\mathbb{R}} g(x)e^{itx} \, F(dx) = 0$$

for all t. Multiply this by a function $\tilde{f}(t) \in L_1(\mathfrak{B}(\mathbb{R}), \ell)$ and integrate with respect to $t \in \mathbb{R}$. Then by Fubini's theorem and the inequality

$$\int_{\mathbb{R}} |g(x)| \, F(dx) \leq \Big(\int_{\mathbb{R}} |g(x)|^2 \, F(dx) \Big)^{1/2} < \infty$$

we obtain

$$\int_{\mathbb{R}} g(x)f(x) \, F(dx) = 0, \quad f(x) := \int_{\mathbb{R}} \tilde{f}(t)e^{itx} \, dt.$$

One knows that every smooth function f with compact support can be written in the above form. Therefore, g is orthogonal to all such functions. The same obviously holds for its real and imaginary parts, which we denote g_r and g_i, respectively. Now for the measures $\mu_{\pm}(dx) = g_{r\pm}(x) \, F(dx)$ we have

$$\int_{\mathbb{R}} f(x) \, \mu_+(dx) = \int_{\mathbb{R}} f(x) \, \mu_-(dx) \tag{2}$$

for every smooth function f with compact support. Then, as in Theorem 1.2.4, we obtain that $\mu_+ = \mu_-$, so that (2) holds for all $f \geq 0$. Substituting $f = g_{r+}$ and noticing that the right-hand side of (2) vanishes, we see that $g_{r+} = 0$ F-almost everywhere. Similarly $g_{r-} = 0$ and $g_{i\pm} = 0$ F-almost everywhere, so that $g = 0$ F-almost everywhere, contradicting the choice of g. The lemma is proved.

2. Theorem (Cramér). *Let ξ_t be a (mean-square continuous) second-order stationary process on $\mathbb{R}$. Let F be the spectral measure of ξ_t. Then, on the collection of all sets of type $(-\infty, a]$, there exists a random orthogonal measure ζ with reference measure F such that $E\zeta(-\infty, a] = 0$ and*

$$\xi_t = \int_{\mathbb{R}} e^{itx} \, \zeta(dx) \quad (a.s.) \quad \forall t. \tag{3}$$

If ζ_1 is another random orthogonal measure having these properties, then, for any $a \in \mathbb{R}$, we have $\zeta_1(-\infty, a] = \zeta(-\infty, a]$ (a.s.).

Proof. Instead of finding ζ in the first place, we will find the stochastic integral against ζ. To do so, define an operator $\Phi : L_2(\mathfrak{B}(\mathbb{R}), F) \to L_2(\mathcal{F}, P)$ in the following way. For f given by (1), define (cf. (3) and Remark 1.10)

$$\Phi f = \sum_{j=1}^{n} c_j \xi_{t_j}.$$

It is easy to check that if f is as in (1), then

$$|f|^2_{L_2(\mathfrak{B}(\mathbb{R}),F)} = E\Big|\sum_{j=1}^{n} c_j \xi_{t_j}\Big|^2 = E|\Phi f|^2, \quad E\Phi f = 0. \tag{4}$$

It follows, in particular, that the operator Φ is well defined on f of type (1) (cf. the argument after Remark 2.3.11). By the way, the fact that it is well defined does not follow from the fact that, if we are given some constants c_j, t_j, c'_j, t'_j, $j = 1, ..., n$, $t_1 < ... < t_n$, $t'_1 < ... < t'_n$, and

$$\sum_{j=1}^{n} c_j e^{it_j x} = \sum_{j=1}^{n} c'_j e^{it'_j x}$$

for *all* x, then the families (c_j, t_j) and (c'_j, t'_j) are the same.

We also see that the operator Φ is a linear isometry defined on the linear subspace of functions (1) as a subspace of $L_2(\mathfrak{B}(\mathbb{R}), F)$ and maps it into $L_2(\mathcal{F}, P)$. By Lemma 2.3.12 it admits a unique extension to an operator defined on the closure in $L_2(\mathfrak{B}(\mathbb{R}), F)$ of this subspace. We keep the notation Φ for this extension and remember that the closure in question coincides with $L_2(\mathfrak{B}(\mathbb{R}), F)$ by Lemma 1. Thus, we have a linear isometric operator $\Phi : L_2(\mathfrak{B}(\mathbb{R}), F) \to L_2(\mathcal{F}, P)$ such that $\Phi e^{it\cdot} = \xi_t$ (a.s.).

Next, observe that $I_{(-\infty,a]} \in L_2(\mathfrak{B}(\mathbb{R}), F)$ and define

$$\zeta(-\infty, a] = \Phi I_{(-\infty,a]}. \tag{5}$$

Since Φ preserves scalar products, we have

$$E\zeta(-\infty, a]\bar{\zeta}(-\infty, b] = F\big((-\infty, a] \cap (-\infty, b]\big).$$

Hence ζ is a random orthogonal measure with reference measure F. Furthermore, it follows from (5) that

$$\Phi f = \int_{\mathbb{R}} f(x)\,\zeta(dx) \tag{6}$$

if f is a step function. Since Φ and the stochastic integral are continuous operators, (6) holds (a.s.) for any $f \in L_2(\mathfrak{B}(\mathbb{R}), F)$. For $f = e^{itx}$ we conclude that

$$\xi_t = \Phi f = \int_{\mathbb{R}} e^{itx}\,\zeta(dx) \quad \text{(a.s.)}.$$

Finally, as has been noticed in (4), we have $E\Phi f = 0$ if f is a function of type (1). For any $f \in L_2(\mathfrak{B}(\mathbb{R}), F)$, take a sequence of functions f_n of type (1) converging to f in $L_2(\mathfrak{B}(\mathbb{R}), F)$ and observe that

$$|E\Phi f| = |E(\Phi f - \Phi f_n)| \leq \big(E|\Phi f - \Phi f_n|^2\big)^{1/2},$$

where the last expression tends to zero by the isometry of Φ and the choice of f_n. Thus, $E\Phi f = 0$ for any $f \in L_2(\mathfrak{B}(\mathbb{R}), F)$. By taking $f = I_{(-\infty,a]}$, we conclude that $E\zeta(-\infty, a] = 0$.

We have proved the "existence" part of our theorem. To prove the uniqueness, define $\Phi_1 f = \int_{\mathbb{R}} f\,\zeta_1(dx)$. The isometric operators Φ and Φ_1 coincide on all functions of type (1), and hence on $L_2(\mathfrak{B}(\mathbb{R}), F)$. In particular, $\zeta(-\infty, a] = \Phi I_{(-\infty,a]} = \Phi_1 I_{(-\infty,a]} = \zeta_1(-\infty, a]$ (a.s.). The theorem is proved.

3. Remark. We have seen in the proof that, for each $f \in L_2(\mathfrak{B}(\mathbb{R}), F)$,

$$E\int_{\mathbb{R}} f\,\zeta(dx) = 0.$$

4. Remark. Let L_2^ξ be the smallest linear closed subspace of $L_2(\mathcal{F}, P)$ containing all $\xi_t, t \in \mathbb{R}$. Obviously the operator Φ in the proof of Theorem 2 is acting from $L_2(\mathfrak{B}(\mathbb{R}), F)$ into L_2^ξ. Therefore, $\zeta(-\infty, a]$ and each integral $\int_{\mathbb{R}} g\,\zeta(dx)$ with $g \in L_2(\mathfrak{B}(\mathbb{R}), F)$ belong to L_2^ξ.

Furthermore, every element of L_2^ξ is representable as $\int_{\mathbb{R}} g\,\zeta(dx)$ with $g \in L_2(\mathfrak{B}(\mathbb{R}), F)$, due to the equality $\xi_t = \int_{\mathbb{R}} \exp(itx)\,\zeta(dx)$ and the isometric property of stochastic integrals.

5. Definition. We say that a complex-valued random vector $(\xi^1, ..., \xi^n)$ is *Gaussian* if for any complex numbers $\lambda_1, ..., \lambda_n$ we have $\sum_j \xi^j \lambda_j = \eta_1 + i\eta_2$, where $\eta = (\eta_1, \eta_2)$ is a two-dimensional Gaussian vector (with real coordinates). As usual, a complex-valued or a real-valued *Gaussian process* is one whose finite-dimensional distributions are all Gaussian.

6. Corollary. *If ξ_t is a Gaussian process, then*

$$\Big(\int_{\mathbb{R}} f_1(x)\,\zeta(dx), ..., \int_{\mathbb{R}} f_n(x)\,\zeta(dx)\Big)$$

is a Gaussian vector for any $f_j \in L_2(\mathfrak{B}(\mathbb{R}), F)$.

This assertion follows from the facts that Φf is Gaussian for trigonometric polynomials and mean-square limits of Gaussian variables are Gaussian.

7. Corollary. *If ξ_t is a* real *valued second-order stationary process, then*

$$\overline{\int_{\mathbb{R}} f(x)\,\zeta(dx)} = \int_{\mathbb{R}} \bar{f}(-x)\,\zeta(dx) \quad (a.s.) \quad \forall f \in L_2(\mathfrak{B}(\mathbb{R}), F).$$

This follows from the fact that the equality holds for $f = e^{itx}$.

8. Exercise. Prove that if both ξ_t and ζ are real valued, then ξ_t is independent of t in the sense that $\xi_t = \xi_s$ (a.s.) for any s, t.

9. Exercise. Let ζ be a random orthogonal measure, defined on all sets $(-\infty, a]$, satisfying $E\zeta(-\infty, a] = 0$ and having finite reference measure. Prove that

$$\int_{\mathbb{R}} e^{itx}\,\zeta(dx)$$

is a mean square continuous second-order stationary process.

10. Definition. The random orthogonal measure whose existence is asserted in Theorem 2 is called *the random spectral measure of ξ_t*, and formula (3) is called *the spectral representation of the process ξ_t*.

For processes with spectral densities which are rational functions, one can give yet another representation of their trajectories. In order to understand how to do this, we start with an important example.

3. Ornstein-Uhlenbeck process

The Wiener process is not second-order stationary because $Ew_t^2 = t$ is not a constant and the distribution of w_t spreads out when time is growing. However, if we add to w_t a drift which would keep the variance moderate, then we can hope to construct a second-order stationary process on the basis of w_t. The simplest way to do so is to consider the following equation:

$$\xi_t = \xi_0 - \alpha \int_0^t \xi_s\,ds + \beta w_t, \tag{1}$$

where α and β are real numbers, $\alpha > 0$, ξ_0 is a real-valued random variable independent of $w_\cdot$, and w_t is a one dimensional Wiener process. For each ω, equation (1) has a unique solution ξ_t defined for all $t \geq 0$, which follows after writing down the equation for $\eta_t := \xi_t - \beta w_t$. Indeed,

$$\dot{\eta}_t = -\alpha(\eta_t + \beta w_t), \quad \eta_0 = \xi_0,$$

$$\eta_t = \xi_0 e^{-\alpha t} - \alpha\beta \int_0^t e^{\alpha(s-t)} w_s \, ds,$$

$$\xi_t = \xi_0 e^{-\alpha t} - \alpha\beta \int_0^t e^{\alpha(s-t)} w_s \, ds + \beta w_t. \tag{2}$$

By Theorem 2.3.22 bearing on integration by parts (cf. Remark 2.4.4), the last formula reads

$$\xi_t = \xi_0 e^{-\alpha t} + \beta \int_0^t e^{\alpha(s-t)} \, dw_s \quad \text{(a.s.)}. \tag{3}$$

1. Theorem. *Let $\xi_0 \sim N(0, \beta^2/(2\alpha))$. Then the solution ξ_t of equation (1) is a Gaussian second-order stationary process on $[0, \infty)$ with zero mean, correlation function*

$$R(t) = \frac{\beta^2}{2\alpha} e^{-\alpha|t|},$$

and spectral density

$$f(x) = \frac{\beta^2}{2\pi} \frac{1}{x^2 + \alpha^2}.$$

Proof. It follows from (3) that $E\xi_t = 0$ and $E|\xi_t|^2 < \infty$. The reader who did Exercise 2.3.23 will understand that the fact that ξ_t is a Gaussian process is proved as in this exercise with the additional observation that, by assumption, ξ_0 is Gaussian and independent of $w_\cdot$.

Next, for $t_1 \geq t_2 \geq 0$, from (3) and the isometric property of stochastic integrals, we get

$$E\xi_{t_1}\xi_{t_2} = \frac{\beta^2}{2\alpha} e^{-\alpha(t_1+t_2)} + \beta^2 \int_0^{t_2} e^{\alpha(2s-t_1-t_2)} \, ds = \frac{\beta^2}{2\alpha} e^{-\alpha(t_2-t_1)}.$$

It follows that ξ_t is second-order stationary with correlation function R. The fact that f is indeed its spectral density is checked by simple computation. The theorem is proved.

2. Definition. A real-valued Gaussian second-order stationary process defined on $\mathbb{R}$ is called an *Ornstein-Uhlenbeck process* if its correlation function satisfies $R(t) = R(0)\exp(-\alpha|t|)$, where α is a nonnegative constant.

3. Exercise. Prove that if ξ_t is a real-valued Gaussian second-order stationary Markov process defined on $\mathbb{R}$, then it is an Ornstein-Uhlenbeck process. Also prove the converse.

Here by the Markov property we mean that

$$E\{f(\xi_t)|\xi_{t_1},...,\xi_{t_n}\} = E\{f(\xi_t)|\xi_{t_n}\} \quad \text{(a.s.)}$$

for any $t_1 \leq ... \leq t_n \leq t$ and Borel f satisfying $E|f(\xi_t)| < \infty$.

Theorem 1 makes it natural to conjecture that any Ornstein-Uhlenbeck process ξ_t should satisfy equation (1) for $t \geq 0$ with *some* Wiener process w_t. To prove the conjecture, for ξ_t satisfying (1), we find w_t in terms of the random spectral measure ζ of ξ_t if $\alpha > 0$ and $\beta > 0$. We need a stochastic version of Fubini's theorem, the proof of which we suggest as an exercise.

4. Exercise*. Let Π be a family of subsets of a set X. Let ζ be a random orthogonal measure defined on Π with reference measure μ defined on $\sigma(\Pi)$. Take a finite interval $[a, b] \subset \mathbb{R}$ and assume that on $[a, b] \times X$ we are given a bounded function $g(t, x)$ which is continuous in $t \in [a, b]$ for any $x \in X$, belongs to $L_2(\Pi, \mu)$ for any $t \in [a, b]$, and satisfies

$$\int_X \sup_{t\in[a,b]} |g(t,x)|^2 \, \mu(dx) < \infty.$$

Prove that $\int_X g(t, x)\, \zeta(dx)$ is continuous in t as an $L_2(\mathcal{F}, P)$-valued function and

$$\int_a^b \Big(\int_X g(t,x)\, \zeta(dx) \Big)\, dt = \int_X \Big(\int_a^b g(t,x)\, dt \Big)\, \zeta(dx),$$

where the first integral against dt is the mean-square integral (see Exercise 1.14) and the second one is the Riemann integral of a continuous function.

By using this result, we find that

$$\beta w_t = \xi_t - \xi_0 + \alpha \int_0^t \xi_s\, ds = \int_{\mathbb{R}} \Big(e^{itx} - 1 + \alpha \int_0^t e^{isx}\, ds \Big)\, \zeta(dx)$$

$$= \int_{\mathbb{R}} \frac{e^{itx} - 1}{ix} (ix + \alpha)\, \zeta(dx),$$

$$w_t = \int_{\mathbb{R}} \frac{e^{itx}-1}{ix} \frac{ix+\alpha}{\beta} \zeta(dx). \tag{4}$$

This representation of w_t will look more natural and invariant if one replaces $\sqrt{2\pi}(ix+\alpha)\beta^{-1}\zeta(dx)$ with a differential of a new random orthogonal measure. To do this rigorously, let $\bar{\Pi} = \{(a,b] : -\infty < a \leq b < \infty\}$ and for $(a,b] \in \bar{\Pi}$ define

$$\lambda(a,b] = \sqrt{2\pi} \int_{\mathbb{R}} I_{(a,b]}(x) \frac{ix+\alpha}{\beta} \zeta(dx).$$

It turns out that λ is a random orthogonal measure with reference measure ℓ. Indeed,

$$E \int_{\mathbb{R}} I_{(a_1,b_1]}(x) \frac{ix+\alpha}{\beta} \zeta(dx) \overline{\int_{\mathbb{R}} I_{(a_2,b_2]}(x) \frac{ix+\alpha}{\beta} \zeta(dx)}$$

$$= \int_{\mathbb{R}} I_{(a_1,b_1]}(x) I_{(a_2,b_2]}(x) \frac{ix+\alpha}{\beta} \frac{-ix+\alpha}{\beta} f(x)\, dx$$

$$= \frac{1}{2\pi} \int_{\mathbb{R}} I_{(a_1,b_1]}(x) I_{(a_2,b_2]}(x)\, dx = \frac{1}{2\pi} \ell((a_1,b_1] \cap (a_2,b_2])$$

(remember that $f = \beta^2(x^2+\alpha^2)^{-1}(2\pi)^{-1}$ and the product of indicators is the indicator of the intersection). By the way, random orthogonal measures with reference measure ℓ are called *standard random orthogonal measures*. Next for any $g \in S(\bar{\Pi}, \ell)$, obviously,

$$\int_{\mathbb{R}} g(x)\, \lambda(dx) = \sqrt{2\pi} \int_{\mathbb{R}} g(x) \frac{ix+\alpha}{\beta} \zeta(dx) \quad \text{(a.s.)}.$$

Actually, this equality holds for any $g \in L_2(\mathfrak{B}(\mathbb{R}), \ell)$, which is proved by standard approximation after noticing that if $g_n \in S(\bar{\Pi}, \ell)$ and $g_n \to g$ in $L_2(\mathfrak{B}(\mathbb{R}), \ell)$, then

$$\int_{\mathbb{R}} \left| g_n(x) \frac{ix+\alpha}{\beta} - g(x) \frac{ix+\alpha}{\beta} \right|^2 f(x)\, dx = \frac{1}{2\pi} \int_{\mathbb{R}} |g_n(x) - g(x)|^2\, dx \to 0.$$

In terms of λ formula (4) takes the form

$$w_t = \frac{1}{\sqrt{2\pi}} \int_{\mathbb{R}} \frac{e^{itx}-1}{ix} \lambda(dx), \quad t \geq 0. \tag{5}$$

Also

$$\xi_t = \frac{1}{\sqrt{2\pi}} \int_{\mathbb{R}} e^{itx} \frac{\beta}{ix+\alpha} \lambda(dx), \quad t \in \mathbb{R}. \tag{6}$$

Now we want to prove that every Ornstein-Uhlenbeck process ξ_t satisfies (1) with the Wiener process w_t *defined* by (5). First of all we need to prove that w_t is indeed a Wiener process. In the future we need a stronger statement, which we prove in the following lemma.

5. Lemma. *Let ξ_t be a* real-valued *Gaussian second-order stationary process defined on $\mathbb{R}$. Assume that it has a spectral density $f(x) \not\equiv 0$ which is represented as $\varphi(x)\bar{\varphi}(x)$, where $\varphi(x)$ is a rational function such that $\bar{\varphi}(x) = \varphi(-x)$ and all poles of $\varphi(z)$ lie in the upper half plane* $\operatorname{Im} z > 0$. *Let ζ be the random spectral measure of ξ_t. For $-\infty < a < b < \infty$ define*

$$\lambda(a,b] = \int_{\mathbb{R}} I_{(a,b]}(x) \frac{1}{\varphi(x)} \zeta(dx) \quad \Big(\frac{1}{\varphi(x)} := 0 \quad \text{if} \quad \varphi(x) = 0\Big),$$

$$w_t = \frac{1}{\sqrt{2\pi}} \int_{\mathbb{R}} \frac{e^{itx}-1}{ix} \lambda(dx), \quad t \geq 0. \tag{7}$$

Then w_t has a continuous modification which is a Wiener process independent of ξ_s, $s \leq 0$.

Proof. Notice that the number of points where $\phi(x) = 0$ is finite and has zero Lebesgue measure. Therefore, in the same way as before the lemma, it is proved that λ is a standard random orthogonal measure, and since $(\exp(itx)-1)/(ix) \in L_2(\mathfrak{B}(\mathbb{R}), \ell)$, the integral in (7) is well defined and

$$w_t = \frac{1}{\sqrt{2\pi}} \int_{\mathbb{R}} \frac{e^{itx}-1}{ix} \frac{1}{\varphi(x)} \zeta(dx).$$

By Corollary 2.6 the process w_t is Gaussian. By virtue of $\bar{\varphi}(-x) = \varphi(x)$ and Corollary 2.7 we get

$$\bar{w}_t = \frac{1}{\sqrt{2\pi}} \int_{\mathbb{R}} \frac{e^{itx}-1}{ix} \frac{1}{\bar{\varphi}(-x)} \zeta(dx) = w_t,$$

so that w_t is real valued. In addition, $w_0 = 0$, $Ew_t = 0$, and

$$Ew_t w_s = Ew_t \bar{w}_s = \frac{1}{2\pi} \int_{\mathbb{R}} \frac{e^{itx}-1}{ix} \frac{e^{-isx}-1}{-ix} dx. \tag{8}$$

One can compute the last integral in two ways. First, if we take ξ_t from Theorem 1, then (4) holds with its left-hand side being a Wiener process by construction. For this process (8) holds, with the first expression known to be $t \wedge s$.

On the other hand, the Fourier transform of $I_{(0,s]}(z)$ is easily computed and turns out to be proportional to $(e^{isx}-1)/(ix)$. Therefore, by Parseval's identity

$$\frac{1}{2\pi}\int_{\mathbb{R}} \frac{e^{itx}-1}{ix}\,\frac{e^{-isx}-1}{-ix}\,dx = \int_{\mathbb{R}} I_{(0,t]}(z)I_{(0,s]}(z)\,dz = t\wedge s.$$

It follows in particular that $E|w_t - w_s|^2 = |t-s|$ and $E|w_t - w_s|^4 = c|t-s|^2$, where c is a constant. By Kolmogorov's theorem, w_t has a continuous modification. This modification, again denoted w_t, is the Wiener process we need.

It only remains to prove that w_t, $t \geq 0$, and ξ_s, $s \leq 0$, are independent. Since $(w_{t_1}, ..., w_{t_n}, \xi_{s_1}, ..., \xi_{s_m})$ is a Gaussian vector for any $t_1, ..., t_n \geq 0$, $s_1, ..., s_m \leq 0$, we need only prove that $Ew_t\xi_s = 0$ for all $t \geq 0 \geq s$. From

$$\xi_s = \int_{\mathbb{R}} e^{isx}\,\zeta(dx) = \int_{\mathbb{R}} e^{isx}\varphi(x)\,\lambda(dx)$$

and (7) we obtain

$$E\xi_s w_t = \frac{1}{\sqrt{2\pi}}\int_{\mathbb{R}} e^{isx}\varphi(x)\,\overline{\Big(\frac{e^{itx}-1}{ix}\Big)}\,dx = \frac{1}{\sqrt{2\pi}}\int_{\mathbb{R}} e^{isx}\varphi(x)\,\frac{e^{-itx}-1}{-ix}\,dx.$$

Remember that $\varphi(z)$ is square integrable over the real line and is a rational function with poles in the upper half plane. Also the functions e^{isz} and e^{-itz} are bounded in the lower half plane. It follows easily that

$$|\varphi(z)| = O\Big(\frac{1}{|z|}\Big), \quad \Big|e^{isz}\varphi(z)\,\frac{e^{-itz}-1}{-iz}\Big| = O\Big(\frac{1}{|z|^2}\Big)$$

for $|z| \to \infty$ with $\operatorname{Im} z \leq 0$. By adding to this the fact that the function

$$e^{isz}\varphi(z)\,\frac{e^{-itz}-1}{-iz}$$

has no poles in the lower half plane, so that by Jordan's lemma its integral over the real line is zero, we conclude that $E\xi_s w_t = 0$. The lemma is proved.

6. Remark. We know that the Wiener process is not differentiable in t. However, especially in technical literature, its derivative, called *the white noise*, is used quite often.

Mathematically speaking, the white noise is a generalized function depending on ω. We want to discuss why it is called "white". There is a complete analogy with white light, which is a mixture of colors corresponding to electromagnetic waves with different frequencies. If one differentiates (7) formally, then

$$\dot{w}_t = \frac{1}{\sqrt{2\pi}} \int_{\mathbb{R}} e^{itx}\, \lambda(dx),$$

which shows that $\dot{w}_t$ is a mixture of all harmonics e^{itx} each taken with the same mean amplitude $(2\pi)^{-1}E|\lambda(dx)|^2 = (2\pi)^{-1}dx$, and the amplitudes corresponding to different frequencies are uncorrelated and moreover independent.

7. Theorem. *Let ξ_t be an Ornstein-Uhlenbeck process with*

$$R(t) = \beta^2(2\alpha)^{-1}e^{-\alpha|t|}, \quad \alpha > 0, \beta > 0.$$

Then, for $t \geq 0$, the process ξ_t admits a continuous modification $\tilde{\xi}_t$ and there exists a Wiener process w_t such that

$$\tilde{\xi}_t = \tilde{\xi}_0 - \alpha \int_0^t \tilde{\xi}_s\, ds + \beta w_t \quad \forall t \geq 0 \tag{9}$$

and w_t, $t \geq 0$, and ξ_s, $s \leq 0$, are independent.

Proof. Define w_t by (4). Obviously Lemma 5 is applicable with $\varphi(x) = \beta(2\pi)^{-1/2}(ix+\alpha)^{-1}$. Therefore, the process w_t has a continuous modification, which is a Wiener process independent of ξ_s, $s \leq 0$, and for which we keep the same notation. Let

$$\tilde{\xi}_t = \xi_0 e^{-\alpha t} - \alpha\beta \int_0^t e^{\alpha(s-t)} w_s\, ds + \beta w_t.$$

By the stochastic Fubini theorem

$$\tilde{\xi}_t = \int_{\mathbb{R}} \left[e^{-\alpha t} - \alpha\beta \int_0^t e^{\alpha(s-t)} \frac{e^{isx}-1}{ix} \frac{ix+\alpha}{\beta}\, ds + \frac{e^{itx}-1}{ix}(ix+\alpha) \right] \zeta(dx)$$

$$= \int_{\mathbb{R}} e^{itx}\, \zeta(dx) = \xi_t$$

(a.s.). In addition $\tilde{\xi}_t$ is continuous and satisfies (9), which is shown by reversing the arguments leading to (2). The theorem is proved.

8. Exercise. We assumed that $\alpha > 0$. Prove that if $\alpha = 0$, then $\xi_t = \xi_0$ (a.s.) for any t.

4. Gaussian stationary processes with rational spectral densities

Let ξ_t be a real-valued Gaussian second-order stationary process. Assume that it has a spectral density $f(x)$ and $f(x) = P_n(x)/P_m(x)$, where P_n and P_m are polynomials of degree n and m respectively. Without loss of generality we assume that P_n and P_m do not have common roots and P_m has the form $x^m + ...$

1. Exercise*. Assume that $f(x) = \tilde{P}_{\tilde{n}}(x)/\tilde{P}_{\tilde{m}}(x)$, where $\tilde{P}_{\tilde{n}}$ and $\tilde{P}_{\tilde{m}}$ do not have common roots and $\tilde{P}_{\tilde{m}}(x) = x^{\tilde{m}} + ...$ Prove that $\tilde{P}_{\tilde{m}}(x) \equiv P_m(x)$ and $\tilde{P}_{\tilde{n}}(x) \equiv P_n(x)$.

Exercise 1 shows that n, m, P_n and P_m are determined uniquely. Moreover, since $\bar{f} = f$, we get that $\bar{P}_n = P_n$ and $\bar{P}_m = P_m$, so that P_n and P_m are real valued. Furthermore, f is summable, so that the denominator P_m does not have real zeros, m is even, and certainly $n < m$. Next, $f \geq 0$ and therefore each real zero of P_n has even multiplicity. Since ξ_t is real valued, by Corollary 1.8, we have $f(x) = f(-x)$, which along with the uniqueness of representation implies that

$$P_n(x) = P_n(-x), \quad P_m(x) = P_m(-x).$$

In turns it follows at once that if a is a root of P_m, then $\bar{a}$, $-\bar{a}$, and $-a$ are also roots of P_m. Remember that m is even, and define

$$Q_+(x) = i^{m/2} \prod_{\operatorname{Im} a_j > 0} (x - a_j), \quad Q_-(x) = i^{-m/2} \prod_{\operatorname{Im} a_j < 0} (x - a_j),$$

where $\{a_j, j = 1, ..., m\}$ are the roots of P_m. Notice that $Q_+(x)Q_-(x) = P_m(x)$ and that, as follows from the above analysis, for real x,

$$\overline{Q_+(x)} = Q_-(x) = i^{-m/2}(-1)^{m/2} \prod_{\operatorname{Im} a_j < 0} (-x - (-a_j))$$

$$= i^{-m/2}(-1)^{m/2} \prod_{\operatorname{Im} a_j > 0} (-x - a_j) = Q_+(-x), \qquad Q_+(x)\overline{Q_+(x)} = P_m(x).$$

Similarly, if P_n does not have real roots, then there exist polynomials P_+ and P_- such that

$$P_-(x) = \overline{P_+(x)} = P_+(-x), \quad P_+(x)\overline{P_+(x)} = P_n(x).$$

Such polynomials exist in the general case as well. In order to prove this, it suffices to notice that, for real a,

$$(x-a)^{2k}(x+a)^{2k} = (x^2-a^2)^k(x^2-a^2)^k, \quad x^{2k} = (ix)^k(-ix)^k.$$

We have proved the following fact with $\varphi = P_+/Q_+$.

2. Lemma. *Let the spectral density $f(x)$ of a real-valued second-order stationary process ξ_t be rational, namely $f(x) = P_n(x)/P_m(x)$, where P_n and P_m are nonnegative polynomials of degree n and m respectively without common roots. Then m is even and $f(x) = \varphi(x)\overline{\varphi(x)}$, where the rational function $\varphi(z)$ has exactly $m/2$ poles all of which lie in the upper half plane and $\bar{\varphi}(x) = \varphi(-x)$ for all $x \in \mathbb{R}$.*

3. Exercise. From the equality $\bar{\varphi}(x) = \varphi(-x)$, valid for all $x \in \mathbb{R}$, derive that $\varphi(ix)$ is real valued for real x.

4. Theorem. *Let the spectral density $f(x)$ of a real-valued Gaussian second-order stationary process ξ_t be a rational function with simple poles. Then there exist an integer $k \geq 1$, (complex) constants α_j and β_j, and continuous Gaussian processes η_t^j and w_t defined for $t \in [0,\infty)$ and $j = 1, ..., k$ such that*

(i) *w_t is a Wiener process, $(\eta_0^1, ..., \eta_0^k)$ is independent of $w_\cdot$, and w_t, $t \geq 0$, is independent of ξ_s, $s \leq 0$;*

(ii) *$\eta_t^j = \eta_0^j - \alpha_j \int_0^t \eta_s^j \, ds + \beta_j w_t$ for any $t \geq 0$;*

(iii) *for $t \geq 0$ we have*

$$\xi_t = \eta_t^1 + ... + \eta_t^k \quad (a.s.).$$

Proof. As in the case of Ornstein-Uhlenbeck processes, we replace the spectral representation $\xi_t = \int_{\mathbb{R}} \exp(itx)\,\zeta(dx)$ with

$$\xi_t = \int_{\mathbb{R}} \varphi(x) e^{itx} \, \lambda(dx),$$

where φ is taken from Lemma 2 and

$$\lambda(a,b] = \int_{\mathbb{R}} I_{(a,b]} \frac{1}{\varphi(x)} \, \zeta(dx) \quad \left(\frac{1}{0} := 0\right).$$

Such replacement is possible owing to the fact that $\varphi\frac{1}{\varphi} = 1$ almost everywhere. It is also seen that λ is a standard orthogonal measure.

Next let

$$\varphi(x) = \frac{\beta_1}{ix+\alpha_1} + ... + \frac{\beta_k}{ix+\alpha_k} \tag{1}$$

be the decomposition of φ into partial fractions. Since the poles of φ lie only in the upper half plane, we have $\operatorname{Re}\alpha_j > 0$. For $t \geq 0$ denote

$$w_t = \frac{1}{\sqrt{2\pi}} \int_{\mathbb{R}} \frac{e^{itx} - 1}{ix} \lambda(dx), \quad \xi_t^j = \int_{\mathbb{R}} \frac{\beta_j}{ix + \alpha_j} e^{itx} \lambda(dx).$$

Observe that ξ_t^j are Gaussian processes by Corollary 2.6 and w_t is a Wiener process by Lemma 3.5. Furthermore, by following our treatment of the Ornstein-Uhlenbeck process one proves existence of a continuous modification η_t^j of ξ_t^j, the independence of $\eta_0 = (\eta_0^1, ..., \eta_0^k)$ and $w_\cdot$, and the fact that

$$\eta_t^j = \eta_0^j - \alpha_j \int_0^t \eta_s^j \, ds + \beta_j w_t, \quad t \geq 0.$$

It only remains to notice that $\xi_t = \xi_t^1 + ... + \xi_t^k = \eta_t^1 + ... + \eta_t^k$ (a.s.). The theorem is proved.

Consider the following system of equations:

$$\begin{cases} \eta_t^1 = \eta_0^1 - \alpha_1 \int_0^t \eta_s^1 \, ds + \beta_1 w_t, \\ \cdots \\ \cdots \\ \eta_t^k = \eta_0^k - \alpha_k \int_0^t \eta_s^k \, ds + \beta_k w_t, \\ \xi_t = \xi_0 - \sum_{j=1}^k \alpha_j \int_0^t \eta_s^j \, ds + w_t \sum_{j=1}^k \beta_j, \end{cases}$$

and the system obtained from it for the real and imaginary parts of η_t^j. Then we get the following result.

5. Theorem. *Under the conditions of Theorem 4, for $t \geq 0$ the process ξ_t has a continuous modification which is represented as the last coordinate of a multidimensional real-valued Gaussian continuous process ζ_t satisfying*

$$\zeta_t = \zeta_0 - \int_0^t A\zeta_s \, ds + w_t B, \quad t \geq 0, \tag{2}$$

where A, B are nonrandom, A is a matrix, B is a vector, w_t is a one-dimensional Wiener process, and ζ_0 and $w_\cdot$ are independent.

6. Remark. Theorem 5 is also true if the multiplicities of the poles are greater than 1. To explain this, observe that then in (1) we also have terms which are constants times higher negative powers of $ix + \alpha_j$, so that we need to understand what kind of equation holds for

$$\kappa_t^n(\alpha) := \int_{\mathbb{R}} \frac{\beta}{(ix+\alpha)^{n+1}} e^{itx} \lambda(dx), \tag{3}$$

where $n \geq 1$, β and α are some complex numbers, and $\operatorname{Re}\alpha > 0$. Arguing formally, one sees that

$$\frac{d^n}{d\alpha^n} \kappa_t^0(\alpha) = (-1)^n n! \, \kappa_t^n(\alpha),$$

and this is the clue. From above we know that there is a continuous modification $\chi_t^0(\alpha)$ of $\kappa_t^0(\alpha)$ satisfying the equation

$$\chi_t^0(\alpha) = \kappa_0^0(\alpha) - \alpha \int_0^t \chi_s^0(\alpha)\, ds + \beta w_t.$$

If we are allowed to differentiate this equation with respect to α, then for

$$\chi_t^j(\alpha) = (-1)^j (j!)^{-1} d^j \chi_t^0(\alpha)/d\alpha^j,$$

after simple manipulations we get

$$\begin{cases} \chi_t^1(\alpha) &= \kappa_0^1(\alpha) - \alpha \int_0^t \chi_s^1(\alpha)\, ds + \int_0^t \chi_s^0(\alpha)\, ds, \\ \ldots \\ \chi_t^n(\alpha) &= \kappa_0^n(\alpha) - \alpha \int_0^t \chi_s^n(\alpha)\, ds + \int_0^t \chi_s^{n-1}(\alpha)\, ds. \end{cases} \tag{4}$$

After having produced (4), we forget the way we did it and derive the result we need rigorously. Define

$$\chi_0^j = \int_{\mathbb{R}} \frac{\beta}{(ix+\alpha)^{j+1}} \lambda(dx)$$

and solve the system

$$\begin{cases} \chi_t^0 &= \chi_0^0 - \alpha \int_0^t \chi_s^0\, ds + \beta w_t, \\ \chi_t^1 &= \chi_0^1 - \alpha \int_0^t \chi_s^1\, ds + \int_0^t \chi_s^0\, ds, \\ \ldots \\ \chi_t^n &= \chi_0^n - \alpha \int_0^t \chi_s^n\, ds + \int_0^t \chi_s^{n-1}\, ds, \end{cases} \tag{5}$$

which is equivalent to a system of first-order linear ordinary differential equations. It turns out that

$$\chi_t^j = \int_{\mathbb{R}} \frac{\beta}{(ix+\alpha)^{j+1}} e^{itx} \lambda(dx) \tag{6}$$

(a.s.) for each $t \geq 0$ and $j = 0, ..., n$. One proves this by induction, noticing that for $j = 0$ this fact is known and, for $j \geq 1$,

$$\chi_t^j = \chi_0^j e^{-\alpha t} + \int_0^t e^{\alpha(s-t)} \chi_s^{j-1} \, ds,$$

so that if (6) holds with $j-1$ in place of j, then, owing to the stochastic Fubini theorem,

$$\chi_t^j = \int_{\mathbb{R}} \left(\frac{\beta}{(ix+\alpha)^j} e^{-\alpha t} + \int_0^t e^{\alpha(s-t)} \frac{\beta}{(ix+\alpha)^{j-1}} e^{isx} \, ds \right) \lambda(dx)$$

$$= \int_{\mathbb{R}} \frac{\beta}{(ix+\alpha)^j} e^{itx} \lambda(dx) \quad \text{(a.s.)}.$$

This completes the induction. Furthermore, χ_0^j and $w.$ are independent, which is proved in the same way as in Lemma 3.5.

Thus, we see that the processes (3) are also representable as the last coordinates of solutions of linear systems of type (5), and the argument proving Theorem 5 works again.

7. Remark. Equation (2) is a multidimensional version of (3.1). In the same way in which we arrived at (3.3), one proves that the solution to (2) is given by

$$\zeta_t = e^{-At}\zeta_0 + \int_0^t e^{A(s-t)} B \, dw_s = e^{-At}\zeta_0 + e^{-At} \int_0^t e^{As} B \, dw_s, \tag{7}$$

where the vector ζ_0 is composed of

$$\eta^{jk} := \int_{\mathbb{R}} \frac{1}{(ix+\alpha_j)^k} \lambda(dx), \tag{8}$$

where $k = 1, ..., n_j$, the $i\alpha_j$'s are the roots of Q_+, and the n_j are their multiplicities.

8. Remark. Similarly to the one-dimensional case, one gives the definition of stationary vector-valued process and, as in Section 3, one proves that the right-hand side of (7) is a Gaussian stationary process even if B is a matrix and w_t is a multidimensional Wiener process, provided that ζ_0 is appropriately distributed and A only has eigenvalues with strictly positive real parts.

9. Remark. We will see later (Sec. 6.11) that solutions of stochastic equations even more complex than (2) have the Markov property, and then we will be able to say that real-valued Gaussian second-order stationary processes with rational spectral density are just components of multidimensional Gaussian Markov processes.

5. Remarks about predicting Gaussian stationary processes with rational spectral densities

We follow the notation from Sec. 4 and again take a real-valued Gaussian second-order stationary process ξ_t and assume that it has a spectral density $f(x)$ which is a rational function. In Sec. 4 we showed that there is a representation of the form $f = |\varphi|^2$ and constructed φ satisfying $\varphi = P_+/Q_+$. Actually, all results of Sec. 4 also hold if we take $\varphi = P_-/Q_+$. It turns out that the choice of $\varphi = P_+/Q_+$ is crucial in applications, in particular, in solving the problem of predicting ξ_t for $t \geq 0$ given observations of ξ_s for $s \leq 0$. We explain this in the series of exercises and remarks below.

1. Exercise. Take $\varphi = P_+/Q_+$. Prove that for each $g \in L_2(\mathfrak{B}(\mathbb{R}), \ell)$,

$$\int_{\mathbb{R}} e^{itx} g(x)\varphi(x)\,dx = 0 \quad \forall t < 0 \Longrightarrow \int_{\mathbb{R}} g(x)\frac{1}{(ix+\alpha_j)^k}\,dx = 0,$$

where $k = 1, ..., n_j$, the $i\alpha_j$'s are the roots of Q_+, and the n_j are their multiplicities.

2. Exercise. Let $L_2^\xi(a,b)$ be the smallest linear closed subspace of $L_2(\mathcal{F}, P)$ containing all $\xi_t, t \in (a,b)$. By using Exercise 1 prove that (see (4.8))

$$\eta^{jk}, \quad \int_{\mathbb{R}} \frac{1}{Q_+(x)}\,\lambda(dx) \in L_2^\xi(-\infty, 0). \tag{1}$$

3. Remark. Now we can explain why we prefer to take P_+/Q_+ and not P_-/Q_+. Here it is convenient to assume that the space $(\Omega, \mathcal{F}, P)$ is complete, so that, if a complete σ-field $\mathcal{G} \subset \mathcal{F}$ and for some functions ζ and η we have $\zeta = \eta$ (a.s.) and ζ is $\mathcal{G}$-measurable, so is η. Let $\mathcal{F}_0^\xi$ be the completion of the σ-field generated by the $\xi_t, t \leq 0$. Notice that in formula (4.7) the random vector ζ_0 is $\mathcal{F}_0^\xi$-measurable by Exercise 2. Owing to the independence of $w_t, t \geq 0$, and $\xi_s, s \leq 0$, for any bounded Borel $h(\zeta)$ (for instance, depending only on the last coordinate of the vector ζ), we have (a.s.)

$$E[h(\zeta_t)|\mathcal{F}_0^\xi] = [Eh(\zeta + \int_0^t e^{A(s-t)} B\, dw_s)]|_{\zeta=\zeta_0 e^{-At}}.$$

We see that now the problem of prediction is reduced to the problem of expressing ζ_0 or equivalently η^{jk} in terms of $\xi_t, t \leq 0$. The following few exercises are aimed at showing how this can be done.

4. Exercise. As a continuation of Exercise 2, prove that if all roots of P_+ are real, then in (1) one can replace $L_2^\xi(-\infty,0)$ with $L_2^\xi(0,\infty)$ or with $L_2^\xi(-\infty,0) \cap L_2^\xi(0,\infty)$. By the way, this intersection can be much richer than only multiples of ξ_0. Consider the case $\varphi(x) = ix/(ix+1)^2$ and prove that

$$\eta := \int_{\mathbb{R}} \frac{1}{(ix+1)^2}\, \lambda(dx) \in L_2^\xi(-\infty,0) \cap L_2^\xi(0,\infty) \quad \text{and} \quad \eta \perp \xi_0.$$

5. Exercise*. We say that a process κ_t, given in a neighborhood of a point t_0, is differentiable *in the mean-square sense at the point* $t = t_0$ and its derivative equals χ if

$$\operatorname*{l.i.m.}_{t \to t_0} \frac{\kappa_t - \kappa_{t_0}}{t - t_0} = \chi.$$

In an obvious way one gives the definition of higher order mean-square derivatives. Prove that ξ_t has $(m-n)/2 - 1$ mean-square derivatives. Furthermore, for $j \leq (m-n)/2 - 1$

$$\frac{d^j}{dt^j}\xi_t = i^j \int_{\mathbb{R}} e^{itx} x^j \varphi(x)\, \lambda(dx),$$

where by d^j/dt^j we mean the jth mean-square derivative.

6. Exercise*. As we have seen in the situation of Exercise 4, $L_2^\xi(-\infty,0) \cap L_2^\xi(0,\infty)$ is not a linear subspace generated by ξ_0. Neither is this a linear subspace generated by ξ_0 and the derivatives of ξ_t at zero, which is seen from the same example given in Exercise 4. In this connection, prove that if $P_+ = \text{const}$, then η^{jk} from (4.8) and ζ_0 do admit representations as values at zero of certain ordinary differential operators applied to ξ_t.

7. Remark. In what concerns ζ_0, the general situation is not too much more complicated than the one described in Exercise 6. Observe that by Exercise 5, the process

$$\tilde{\xi}_t = \int_{\mathbb{R}} \frac{1}{Q_+(x)}\, e^{itx}\, \lambda(dx)$$

satisfies the equation $P_+(-iD_t)\tilde{\xi}_t = \xi_t$, $t \leq 0$, which is understood as an equation for L_2-valued functions with appropriate conditions at $-\infty$ (cf. the hint to Exercise 1 and Exercise 8). The theory of ordinary differential equations for Banach-space-valued functions is quite parallel to that of real-valued functions. In particular, well known formulas for solutions of linear equations are available. Therefore, there are formulas expressing $\tilde{\xi}_t$ through ξ_s, $s \leq 0$. Furthermore, as in Exercise 6 the random variables η^{jk} from (4.8) are representable as values at zero of certain ordinary differential operators applied to $\tilde{\xi}_t$.

8. Exercise. For $\varepsilon > 0$, let $P_+^{\varepsilon}(x) = P_+(x - i\varepsilon)$. Prove that there exists a unique solution of $P_+^{\varepsilon}(-iD_t)\tilde{\xi}_t^{\varepsilon} = \xi_t$ on $(-\infty, 0)$ in the class of functions $\tilde{\xi}_t^{\varepsilon}$ for which $E|\tilde{\xi}_t^{\varepsilon}|^2$ is bounded. Also prove that $\text{l.i.m.}_{\varepsilon\downarrow 0}\, \tilde{\xi}_t^{\varepsilon} = \tilde{\xi}_t$.

6. Stationary processes and the Birkhoff-Khinchin theorem

For second-order stationary processes the covariance between ξ_t and ξ_{t+s} is independent of the time shift. There are processes possessing stronger time shift invariance properties.

1. Definition. A real-valued process ξ_n given for integers $n \in (T, \infty)$ is said to be *stationary* if for any integers $k_1, ..., k_n \in (T, \infty)$ and $i \geq 0$ the distributions of the vectors $(\xi_{k_1}, ..., \xi_{k_n})$ and $(\xi_{k_1+i}, ..., \xi_{k_n+i})$ coincide.

Usually we assume that $T = -1$, so that ξ_n is given for $n = 0, 1, 2, ...$ Observe that, obviously, if ξ_n is stationary, then $f(\xi_n)$ is stationary for any Borel f.

2. Exercise*. Prove that ξ_n, $n = 0, 1, 2, ...$, is stationary if and only if for each integer n, the vectors $(\xi_0, ..., \xi_n)$ and $(\xi_1, ..., \xi_{n+1})$ have the same distribution.

3. Example. The sequence $\xi_n \equiv \eta$ is stationary for any random variable η.

4. Example. Any sequence of independent identically distributed random variables is stationary.

5. Example. This example generalizes both Examples 3 and 4. Remember that a random sequence $\xi_0, \xi_1, ...$ is called *exchangeable* if for every n and every permutation π of $\{0, 1, 2, ..., n\}$, the distribution of $(\xi_{\pi(0)}, ..., \xi_{\pi(n)})$ coincides with that of $(\xi_0, ..., \xi_n)$.

It turns out that if a sequence $\xi_0, \xi_1, ...$ is exchangeable, then it is stationary. Indeed, for any Borel bounded f

$$Ef(\xi_0, ..., \xi_{n+1}) = Ef(\xi_1, ..., \xi_{n+1}, \xi_0).$$

By taking f independent of the last coordinate, we see that the laws of the vectors $(\xi_1, ..., \xi_{n+1})$ and $(\xi_0, ..., \xi_n)$ coincide, so that ξ_n is stationary by Exercise 2.

6. Example. Clearly, if $\xi_0, \xi_1, ...$ is stationary and $E|\xi_k|^2 < \infty$ for some k, the same holds for any $k > T$ and $E\xi_n\xi_k$ does not change under the translations $n \to n+i$, $k \to k+i$. Therefore, $E\xi_n\xi_k$ depends only on the difference $k-n$, and ξ_n is a mean-square stationary process (sequence). The converse is also true if ξ_n is a Gaussian sequence, since then the finite-dimensional distributions of $\xi_\cdot$, which are uniquely determined by the mean value and the covariance function, do not change under translations of time. In particular, the Ornstein-Uhlenbeck process (considered at integral times) is stationary.

7. Example. Let Ω be a circle of length 1 centered at zero with Borel σ-field and linear Lebesgue measure. Fix a point $x_0 \in \Omega$ and think of any other point $x \in \Omega$ as the length of the arc from x_0 to x in the clockwise direction. Then the operation $x_1 + x_2$ is well defined. Fix $\alpha \in \Omega$ and define $\xi_n(\omega) = \omega + n\alpha$. Since the distribution of $\omega + x$ is the same as that of ω for any x, we have that the distribution of $(\xi_0(\omega + x), ..., \xi_n(\omega + x))$ coincides with that of $(\xi_0(\omega), ..., \xi_n(\omega))$ for any x. By taking $x = \alpha$, we conclude that ξ_n is a stationary process.

For stationary processes we will prove only one theorem, namely the Birkhoff-Khinchin theorem. This theorem was first proved by Birkhoff, then generalized by Khinchin. Kolmogorov, F. Riesz, E. Hopf and many others invented various proofs and generalizations of the theorem. All these proofs, however, were quite involved. Only at the end of the sixties did Garsia find an elementary proof of the key Hopf inequality which made it possible to present the proof of the Birkhoff-Khinchin theorem in this introductory book.

The proof, given below, consists of two parts, the first being the proof of the Hopf maximal inequality, and the second being some more or less general manipulations. In order to get acquainted with these manipulations, we show them first not for stationary processes but for reverse martingales. We will see again that they have (a.s.) limits as $n \to \infty$, this time without using Doob's upcrossing theorem.

Remember that a sequence $(\eta_n, \mathcal{F}_n)$ is called a reverse martingale if the σ-fields $\mathcal{F}_n \subset \mathcal{F}$ decrease in n, η_n is $\mathcal{F}_n$-measurable, $E|\eta_n| < \infty$ and

$$E\{\eta_n | \mathcal{F}_{n+1}\} = \eta_{n+1} \quad \text{(a.s.)}.$$

Then $\eta_n = E\{\eta_0|\mathcal{F}_n\}$ and, as we know (Theorem 3.4.9), the limit of η_n exists almost surely as $n \to \infty$.

Let us prove this fact starting with the Kolmogorov-Doob inequality: for any $p \in \mathbb{R}$ (and not only $p > 0$),

$$E\eta_0 I_{\max_{i\le n}\eta_i > p} = \sum_{i=0}^{n-1} E\eta_0 I_{\eta_n,\dots,\eta_{n-i}\le p,\eta_{n-i-1}>p} + E\eta_0 I_{\eta_n>p}$$

$$= \sum_{i=0}^{n-1} E\eta_{n-i-1} I_{\eta_n,\dots,\eta_{n-i}\le p,\eta_{n-i-1}>p} + E\eta_n I_{\eta_n>p}$$

$$\ge p\left(\sum_{i=0}^{n-1} P(\eta_n,\dots,\eta_{n-i}\le p,\eta_{n-i-1}>p) + P(\eta_n>p)\right) = pP(\max_{i\le n}\eta_i > p).$$

From the above proof it is also seen that if $A \in \mathcal{F}_\infty := \bigcap_n \mathcal{F}_n$, then

$$E\eta_0 I_{A,\max_{i\le n}\eta_i>p} \ge pP(A, \max_{i\le n}\eta_i > p). \tag{1}$$

Take here

$$A = B\cap C_p, \quad C_p := \{\omega : \varlimsup_{n\to\infty} \eta_n > p\}, \quad B \in \mathcal{F}_\infty.$$

Clearly, for $n \ge n_0$, the random variable $\sup_{i\ge n}\eta_i$ is $\mathcal{F}_n$- and $\mathcal{F}_{n_0}$-measurable. Hence, $C_p \in \mathcal{F}_{n_0}$ and $C_p \in \mathcal{F}_\infty$. Furthermore, $C_p \cap \{\max_{i\le n}\eta_i > p\} \uparrow C_p$ as $n \to \infty$. Therefore, employing also the dominated convergence theorem, from (1), we get

$$E\eta_0 I_{B\cap C_p,\max_{i\le n}\eta_i>p} \ge pP(B\cap C_p, \max_{i\le n}\eta_i > p), \tag{2}$$

$$E\eta_0 I_{B\cap C_p} \ge pP(B\cap C_p). \tag{3}$$

By replacing η_n with $-\eta_n$ and p with $-p$, for any $q \in \mathbb{R}$, we obtain

$$E\eta_0 I_{B\cap D_q} \le qP(B\cap D_q) \quad \text{with} \quad D_q := \{\omega : \varliminf_{n\to\infty} \eta_n < q\}. \tag{4}$$

Now take $B = D_q$ in (3) and $B = C_p$ in (4). Then

$$pP(\underline{\lim}_{n\to\infty} \eta_n < q, \overline{\lim}_{n\to\infty} \eta_n > p) \le E\eta_0 I_{D_q\cap C_p} \le qP(\underline{\lim}_{n\to\infty} \eta_n < q, \overline{\lim}_{n\to\infty} \eta_n > p).$$

For $p > q$, these inequalities are only possible if

$$P(\underline{\lim}_{n\to\infty} \eta_n < q, \overline{\lim}_{n\to\infty} \eta_n > p) = 0.$$

Therefore, the set

$$\{\omega : \underline{\lim}_{n\to\infty} \eta_n < \overline{\lim}_{n\to\infty} \eta_n\} = \bigcup_{\substack{\text{rational } p,q \\ p>q}} \{\omega : \underline{\lim}_{n\to\infty} \eta_n < q, \overline{\lim}_{n\to\infty} \eta_n > p\}$$

has probability zero, and this proves that $\lim_{n\to\infty} \eta_n$ exists almost surely.

Coming back to stationary processes, we give the following definition.

8. Definition. An event A is called *invariant* if for each $n \ge 0$ and Borel $f(x_0, ..., x_n)$ such that $E|f(\xi_0, ..., \xi_n)| < \infty$, we have

$$Ef(\xi_1, ..., \xi_{n+1})I_A = Ef(\xi_0, ..., \xi_n)I_A.$$

Denote

$$S_n = \xi_0 + ... + \xi_n \quad n \ge 0, \quad \bar{l} = \overline{\lim}_{n\to\infty} \frac{S_n}{n}, \quad \underline{l} = \underline{\lim}_{n\to\infty} \frac{S_n}{n}.$$

9. Lemma. *For any Borel $B \subset \mathbb{R}^2$, the event $\{\omega : (\bar{l}, \underline{l}) \in B\}$ is invariant.*

Proof. Fix $n \ge 0$ and without loss of generality only concentrate on Borel bounded $f \ge 0$. Define

$$\mu_i(B) = Ef(\xi_i, ..., \xi_{n+i})I_{(\bar{l},\underline{l})\in B}.$$

We need to prove that $\mu_0(B) = \mu_1(B)$. Since the μ_i's are finite measures on Borel B's, it suffices to prove that the integrals of bounded continuous functions against μ_i's coincide. Let g be such a function. Then

$$\int_{\mathbb{R}^2} g(x, y)\, \mu_i(dxdy) = Ef(\xi_i, ..., \xi_{n+i})g(\bar{l}, \underline{l}).$$

Next, let $S'_n = \xi_1 + ... + \xi_{n+1}$. Then, by using the stationarity of ξ_k and the dominated convergence theorem and denoting $F_i = f(\xi_i, ..., \xi_{n+i})$, we find that

$$EF_0 g(\bar{l}, \underline{l}) = \lim_{k_1\to\infty} EF_0 g(\bar{l}, \inf_{r\ge k_1} [S_r/(r+1)])$$

$$= \lim_{k_1\to\infty} \lim_{k_2\to\infty} EF_0 g(\bar{l}, \min_{k_2\ge r\ge k_1} [S_r/(r+1)])$$

$$= \lim_{k_1\to\infty}\lim_{k_2\to\infty}\lim_{k_3\to\infty}\lim_{k_4\to\infty} EF_0 g(\max_{k_4\geq r\geq k_3}[S_r/(r+1)], \min_{k_2\geq r\geq k_1}[S_r/(r+1)])$$

$$= EF_1 g(\overline{\lim_{n\to\infty}}[S'_n/(n+1)], \underline{\lim}_{n\to\infty}[S'_n/(n+1)]) = EF_1 g(\bar{l}, \underline{l}).$$

The lemma is proved.

Now comes the key lemma.

10. Lemma (Hopf). *Let A be an invariant event and $E|\xi_0| < \infty$. Then for all $p \in \mathbb{R}$ and $n = 1, 2, ...$, we have*

$$E\xi_0 I_{A,\max_{0\leq i\leq n}[S_i/(i+1)]>p} \geq pP\{A, \max_{0\leq i\leq n}[S_i/(i+1)] > p\}.$$

Proof (Garsia). First assume $p = 0$ and use the obvious equality

$$\max_{0\leq i\leq n} S_i = \xi_0 + \max\{0, S^1_1, ..., S^1_n\} = \xi_0 + (\max_{1\leq i\leq n} S^1_i)_+,$$

where $S^1_n = \xi_1 + ... + \xi_n$. Also notice that

$$E\max_{0\leq i\leq n}|S_i| \leq E(|\xi_0| + ... + |\xi_n|) = (n+1)E|\xi_0| < \infty.$$

Then, for any invariant A,

$$E\xi_0 I_{A,\max_{0\leq i\leq n}[S_i/(i+1)]>0} = E\xi_0 I_{A,\max_{0\leq i\leq n} S_i>0}$$

$$= E(\max_{0\leq i\leq n} S_i) I_{A,\max_{0\leq i\leq n} S_i>0} - E(\max_{1\leq i\leq n} S^1_i)_+ I_{A,\max_{0\leq i\leq n} S_i>0}$$

$$\geq E(\max_{0\leq i\leq n} S_i)_+ I_A - E(\max_{1\leq i\leq n+1} S^1_i)_+ I_A.$$

The last expression is zero by definition, since A is invariant.

This proves the lemma for $p = 0$. In the general case, it suffices to consider $\xi_i - p$ instead of ξ_i and notice that $S_i/(i+1) > p$ if and only if $(\xi_0 - p) + ... + (\xi_i - p) > 0$. The lemma is proved.

11. Theorem (Birkhoff-Khinchin). *Let ξ_n be a stationary process and $f(x)$ a Borel function such that $E|f(\xi_0)| < \infty$. Then* (i) *the limit*

$$f^* := \lim_{n\to\infty}\frac{1}{n+1}[f(\xi_0) + ... + f(\xi_n)]$$

exists almost surely, and (ii) *we have*

$$E|f^*| \leq E|f(\xi_0)|, \quad \lim_{n\to\infty} E\Big|f^* - \frac{1}{n+1}[f(\xi_0) + ... + f(\xi_n)]\Big| = 0. \quad (5)$$

Proof. (i) Since $f(\xi_n)$ is a stationary process, without loss of generality we may and will take $f(\xi_n) = \xi_n$ and assume that $E|\xi_0| < \infty$. In this situation we just repeat almost word for word the above proof of convergence for reverse martingales.

Denote $\eta_n = S_n/(n+1)$. Then Hopf's lemma says that (2) holds provided $B \cap C_p$ is invariant. By letting $n \to \infty$ we obtain (3). Changing signs leads to (4) provided $B \cap D_q$ is invariant. Lemma 9 allows us to take $B = D_q$ in (3) and $B = C_p$ in (4). The rest is exactly the same as above, and assertion (i) follows.

(ii) The first equation in (5) follows from (i), Fatou's lemma, and the fact that $E|f(\xi_k)| = E|f(\xi_0)|$. The second one follows from (i) and the dominated convergence theorem if f is bounded. In the general case, take any $\varepsilon > 0$ and find a bounded Borel g such that $E|f(\xi_0) - g(\xi_0)| \leq \varepsilon$. Then

$$\overline{\lim_{n\to\infty}} E\Big|f^* - \frac{1}{n+1}[f(\xi_0) + ... + f(\xi_n)]\Big|$$

$$\leq \overline{\lim_{n\to\infty}} E\Big|f^* - g^* - \frac{1}{n+1}[\{f(\xi_0) - g(\xi_0)\} + ... + \{f(\xi_n) - g(\xi_n)\}]\Big|$$

$$\leq E|(f-g)^*| + E|f(\xi_0) - g(\xi_0)| \leq 2E|f(\xi_0) - g(\xi_0)| \leq 2\varepsilon.$$

Since $\varepsilon > 0$ is arbitrary, we get the second equation in (5). The theorem is proved.

12. Exercise. We concentrated on real-valued stationary processes only for the sake of convenience of notation. One can consider stationary processes with values in arbitrary measure spaces, and the Birkhoff-Khinchin theorem with its proof carries over to them without any change. Moreover, obviously instead of real-valued f one can take $\mathbb{R}^d$-valued functions. In connection with this, prove that if ξ_n is a (real-valued) stationary process, f is a Borel function satisfying $E|f(\xi_0)| < \infty$, and z is a complex number with $|z| = 1$, then the limit

$$\lim_{n\to\infty} \frac{1}{n+1}[z^0 f(\xi_0) + ... + z^n f(\xi_n)]$$

exists almost surely.

The Birkhoff-Khinchin theorem looks like the strong law of large numbers, and its assertion is most valuable when the limit is nonrandom. In that case (5) implies that $f^* = Ef(\xi_0)$. In other words,

$$\lim_{n\to\infty} \frac{1}{n+1}[f(\xi_0) + ... + f(\xi_n)] = Ef(\xi_0) \quad \text{(a.s.)}. \tag{6}$$

Let us give some conditions for the limit to be constant.

13. Definition. A stationary process ξ_n is said to be *ergodic* if any invariant event A belonging to $\sigma(\xi_0, \xi_1, ...)$ has probability zero or one.

14. Theorem. *If ξ_n is a stationary ergodic process, then* (6) *holds for any Borel function f satisfying $E|f(\xi_0)| < \infty$.*

Proof. By Lemma 9, for any constant c, the event $\{f^* \leq c\}$ is invariant and, obviously, belongs to $\sigma(\xi_0, \xi_1, ...)$. Because of ergodicity, $P(f^* \leq c) = 0$ or 1. Since this holds for any constant c, $f^* =$ const (a.s.) and, as we have seen before the theorem, (6) holds indeed. The theorem is proved.

The Birkhoff-Khinchin theorem for ergodic processes is important in physics. For instance, take the problem of finding the average magnitude of the speed of molecules of a gas in a given volume. Assume that the gas is in a stationary regime, so that, in particular, this average is independent of time. It is absolutely impossible to measure the speeds of all molecules at a given time and then compute the average in question. The Birkhoff-Khinchin theorem guarantees, on the intuitive level, that if we take "almost any" particular molecule and measure its speed at moments $0, 1, 2, ...$, then the arithmetic means of magnitudes of these measurements will converge to the average magnitude of speed of all molecules. Physical intuition tells us that in order for this to be true, the molecules of gas should intermix "well" during their displacements. In mathematical terms this translates to the requirement of ergodicity. We may say that if there is a good mixing or ergodicity, then the individual average over time coincides with the average over the ensemble of all molecules.

Generally, stationary processes need not be ergodic, as it is seen from Example 3. For many of those that are ergodic, proving ergodicity turns out to be very hard. On the other hand, there are some cases in which checking ergodicity is rather simple.

15. Theorem. *Any sequence of i.i.d. random variables is ergodic.*

Proof. Let ξ_n be a sequence of i.i.d. random variables and let A be an invariant event belonging to $\sigma := \sigma(\xi_0, \xi_1, ...)$. Define $\Pi = \bigcup_n \sigma(\xi_0, \xi_1, ...\xi_n)$. Then, for each n and Borel Γ, we have

$$\{\omega : \xi_n \in \Gamma\} \in \sigma(\xi_0, \xi_1, ...\xi_n) \subset \Pi \subset \sigma(\Pi),$$

so that $\sigma \subset \sigma(\Pi)$. On the other hand, $\Pi \subset \sigma$ and $\sigma(\Pi) \subset \sigma$. Thus $\sigma(\Pi) = \sigma$, which by Theorem 2.3.19 implies that $L_1(\sigma, P) = L_1(\Pi, P)$. In particular, for any $\varepsilon \in (0, 1)$, there are an n and a $\sigma(\xi_0, \xi_1, ...\xi_n)$-measurable random variable f such that

$$E|I_A - f| \le \varepsilon.$$

Without loss of generality, we may assume that $|f| \le 2$ and that f takes only finitely many values.

Next, by using the fact that any element of $\sigma(\xi_0, \xi_1, ...\xi_n)$ has the form $\{\omega : (\xi_0, \xi_1, ...\xi_n) \in B\}$, where B is an appropriate Borel set in $\mathbb{R}^{n+1}$, it is easy to prove that $f = f(\xi_0, ..., \xi_n)$, where $f(x_0, ..., x_n)$ is a Borel function. Therefore, the above assumptions imply that

$$P(A) = EI_AI_A \le \varepsilon + Ef(\xi_0, ..., \xi_n)I_A$$

$$= \varepsilon + Ef(\xi_{n+1}, ..., \xi_{2n+1})I_A \le 3\varepsilon + Ef(\xi_{n+1}, ..., \xi_{2n+1})f(\xi_0, ..., \xi_n)$$

$$= 3\varepsilon + [Ef(\xi_0, ..., \xi_n)]^2 \le 3\varepsilon + [P(A) + \varepsilon]^2.$$

By letting $\varepsilon \downarrow 0$, we conclude that $P(A) \le [P(A)]^2$, and our assertion follows. The theorem is proved.

From this theorem and the Birkhoff-Khinchin theorem we get Kolmogorov's strong law of large numbers for i.i.d. random variables with $E|\xi_0| < \infty$. This theorem also allows one to get stronger results even for the case of ξ_n which are i.i.d. As an example let $\eta_n = f(\xi_n, \xi_{n+1}, ...)$, where f is independent of n. Assume $E|\eta_0| < \infty$. Then η_n is a stationary process and the event

$$\{\omega : \lim_{n\to\infty} \frac{1}{n+1}(\eta_0 + ... + \eta_n) < c\}$$

is invariant *with respect to the process* ξ_n. Therefore, the limit is constant with probability 1. As above, one proves that the limit equals $E\eta_0$ (a.s.).

In Example 7, for α irrational, one could also prove that ξ_n is an ergodic process (see Exercise 16), and this would lead to (1.2.7) for almost every x. Notice that in Exercise 1.2.13 we have already seen that actually (1.2.7) holds for *any* x provided that f is Borel and Riemann integrable. The application of the Birkhoff-Khinchin theorem allows one to extend this result for any Borel function that is integrable with convergence for almost all x.

16. Exercise. Prove that the process from Example 7 is ergodic if α is irrational.

Finally, let us prove that if ξ_n is a real-valued Gaussian second order stationary process with correlation function tending to zero at infinity, then $f^* = Ef(\xi_0)$ (a.s.) for any Borel f such that $E|f(\xi_0)| < \infty$. By the way, f^* exists due to Example 6 and the Birkhoff-Khinchin theorem.

Furthermore, owing to the first relation in (5), to prove $f^* = Ef(\xi_0)$ it suffices to concentrate on bounded and uniformly continuous f. In that case, $g(\xi_n) := f(\xi_n) - Ef(\xi_0)$ is a second order stationary process. As in

Exercise 1.14 (actually easier because there is no need to use mean-square integrals) one proves that

$$\text{l.i.m.}\ \frac{1}{n+1}(g(\xi_0)+...+g(\xi_n))=0 \tag{7}$$

if

$$\lim_{n\to\infty} Eg(\xi_0)g(\xi_n)\to 0. \tag{8}$$

By the Birkhoff-Khinchin theorem, the limit in (7) exists pointwise (a.s.) and it coincides, of course, with the mean-square limit. It follows that we need only prove (8).

Without loss of generality assume $R(0)=1$. Then $\eta_n:=\xi_n-R(n)\xi_0$ and ξ_0 are uncorrelated and hence independent. By using this and the uniform continuity of g we conclude that

$$\lim_{n\to\infty} Eg(\xi_0)g(\xi_n)=\lim_{n\to\infty} Eg(\xi_0)g(\eta_n+R(n)\xi_0)$$

$$=\lim_{n\to\infty} Eg(\xi_0)g(\eta_n)=\lim_{n\to\infty} Eg(\xi_0)Eg(\eta_n)=0,$$

the last equality being true because $Eg(\xi_0)=0$.

7. Hints to exercises

1.11 Instead of Fourier integrals, consider Fourier series.

1.14 Use that continuous H-valued functions are uniformly continuous.

1.15 Observe that our assertions can be expressed in terms of R only, since, for every continuous nonrandom f,

$$E\Big|\int_a^b \xi_t f_t\,dt\Big|^2=E\Big|\int_a^b \eta_t f_t\,dt\Big|^2$$

whenever ξ_t and η_t have the same correlation function. Another useful observation is that, if $R(0)=1$, then $R(t)=Ee^{it\xi}=F\{0\}+Ee^{it\xi}I_{\xi\neq 0}$, and

$$\frac{1}{T}\int_0^T R(t)\,dt=F\{0\}+EI_{\xi\neq 0}[e^{iT\xi}-1]/(iT\xi).$$

3.3 In the proof of the converse, notice that, if $R(0)=1$, then ξ_r and $\xi_{t+s}-e^{-\alpha s}\xi_t$ are uncorrelated, hence independent for $r\le t$, $s\ge 0$.

3.4 Write the left-hand side as the mean-square limit of integral sums, and use the isometric property of the stochastic integral along with the dominated convergence theorem to find the L_2-limit.

4.1 From $P_m(x)\tilde{P}_{\tilde{n}}(x) \equiv \tilde{P}_{\tilde{m}}(x)P_n(x)$ conclude that any root of P_m is a root of $\tilde{P}_{\tilde{m}}$, but not of P_n since P_m and P_n do not have common roots. Then derive that $\tilde{P}_{\tilde{m}}(x) \equiv P_m(x)$.

4.3 Observe that $\bar{\varphi}(x)|_{x=z} = \varphi(-x)|_{x=z}$ for all complex z and $\bar{\varphi}(x)|_{x=-iy} = \overline{\varphi(iy)}$ for real y.

5.1 Define

$$G(t) = \int_{\mathbb{R}} e^{itx} g(x) \frac{1}{Q_+(x)}\, dx$$

and prove that G is $m/2 - 1$ times continuously differentiable in t and tends to zero as $|t| \to \infty$ as the Fourier transform of an L_1 function. Then prove that G satisfies the equation $P_+(-iD_t)G(t) = 0$ for $t \leq 0$, where $D_t = d/dt$. Solutions of this linear equation are linear combinations of some integral powers of t times exponential functions. Owing to the choice of P_+, its roots lie in the closed upper half plane, which implies that the exponential functions are of type $\exp(at)$ with $\operatorname{Re} a \leq 0$, none of which goes to zero as $t \to -\infty$. Since $G(t) \to 0$ as $t \to -\infty$, we get that $G(t) = 0$ for $t \leq 0$. Now apply linear differential operators to G to get the conclusion.

5.2 Remember the definition of L_2^ξ from Remark 2.4. By this remark, if $\eta \in L_2^\xi$, then $\eta = \int_{\mathbb{R}} g(x)\,\lambda(dx)$ with $g \in L_2(\mathfrak{B}(\mathbb{R}), \ell)$. If in addition $\eta \perp L_2^\xi(-\infty, 0)$, then $\int_{\mathbb{R}} \bar{g}(x)e^{itx}\varphi(x)\,dx = 0$ for $t \leq 0$. Exercise 5.1 shows then that η is orthogonal to the random variables in (5.1).

5.8 For the uniqueness see the hint to Exercise 5.1. Also notice that P_+^ε does not have real roots, and

$$\tilde{\xi}_t^\varepsilon = \int_{\mathbb{R}} \frac{P_+(x)}{P_+^\varepsilon(x)Q_+(x)}\, e^{itx}\, \lambda(dx).$$

6.2 If the distributions of two vectors coincide, the distributions of their respective subvectors coincide too. Therefore, for any $i \leq n$, the vectors $(\xi_i, ..., \xi_n)$ and $(\xi_{i+1}, ..., \xi_{n+1})$ have the same distribution.

6.12 Notice that the process $\eta_n := z^n e^{i\omega}$, where ω is $\Omega = [0, 2\pi]$ with Lebesgue measure, is stationary. Also notice that the product of two independent stationary processes is stationary.

6.16 For an invariant set A and any integers $m \in \mathbb{R}$ and $k \geq 0$ we have

$$\int_{\Omega} e^{2\pi i m\omega} I_A(\omega)\, d\omega = e^{2\pi i m\alpha} \int_{\Omega} e^{2\pi i m\omega} I_A(\omega)\, d\omega$$

$$= e^{2\pi imk\alpha} \int_{\Omega} e^{2\pi im\omega} I_A(\omega) \, d\omega,$$

where $d\omega$ is the differential of the linear Lebesgue measure and k is any integer. By using (1.2.6), conclude that, for any square-integrable random variable f, $EfI_A = P(A)Ef$. Then take $f = I_A$.

Chapter 5

Infinitely Divisible Processes

The Wiener process has independent increments and the distribution of each increment depends only on the length of the time interval over which the increment is taken. There are many other processes possessing this property; for instance, the Poisson process or the process τ_a, $a \geq 0$, from Example 2.5.8 are examples of those (see Theorem 2.6.1).

In this chapter we study what can be said about general processes of that kind. They are supposed to be given on a *complete* probability space $(\Omega, \mathcal{F}, P)$ usually behind the scene. The assumption that this space is complete will turn out to be convenient to use starting with Exercise 5.5. One more stipulation is that unless explicitely stated otherwise, all the processes under consideration are assumed to be *real valued*. Finally, after Theorem 1.5 we tacitly assume that all processes under consideration are stochastically continuous without specifying this each time.

1. Stochastically continuous processes with independent increments

We start with processes having independent increments. The main goal of this section is to show that these processes, or at least their modifications, have rather regular trajectories (see Theorem 11).

1. Definition. A real- or vector-valued random process ξ_t given on $[0,\infty)$ is said to be a process *with independent increments* if $\xi_0 = 0$ (a.s.) and $\xi_{t_1}, \xi_{t_2} - \xi_{t_1}, ..., \xi_{t_n} - \xi_{t_{n-1}}$ are independent provided $0 \leq t_1 \leq ... \leq t_n < \infty$.

We will be only dealing with stochastically continuous processes.

2. Definition. A real- or vector-valued random process ξ_t given on $[0,\infty)$ is said to be *stochastically continuous* at a point $t_0 \in [0,\infty)$ if $\xi_t \xrightarrow{P} \xi_{t_0}$ as $t \to t_0$. We say that ξ_t is stochastically continuous on a set if it is stochastically continuous at each point of the set.

Clearly, ξ_t is stochastically continuous at t_0 if $E|\xi_t - \xi_{t_0}| \to 0$ as $t \to t_0$. Stochastic continuity is very weakly related to the continuity of trajectories. For instance, for the Poisson process with parameter 1 (see Exercise 2.3.8) we have $E|\xi_t - \xi_{t_0}| = |t - t_0|$. However, all trajectories of ξ_t are discontinuous. By the way, this example shows also that the requirement $\beta > 0$ in Kolmogorov's Theorem 1.4.8 is essential. The trajectories of τ_a, $a \geq 0$, are also discontinuous, but this process is stochastically continuous too since (see Theorem 2.6.1 and (2.5.1))

$$P(|\tau_b - \tau_a| > \varepsilon) = P(\tau_{|b-a|} > \varepsilon) = P(\max_{t \leq \varepsilon} w_s < |b-a|) \to 0 \quad \text{as} \quad b \to a.$$

3. Exercise. Prove that, for any ω, the function τ_a, $a > 0$, is *left continuous* in a.

4. Definition. A (real-valued) random process ξ_t given on $[0,\infty)$ is said to be *bounded in probability* on a set $I \subset [0,\infty)$ if

$$\lim_{c \to \infty} \sup_{t \in I} P(|\xi_t| > c) = 0.$$

As in usual analysis, one proves the following.

5. Theorem. *If the process ξ_t is stochastically continuous on $[0,T]$ ($T < \infty$), then*

(i) *it is uniformly stochastically continuous on $[0,T]$, that is, for any $\gamma, \varepsilon > 0$ there exists $\delta > 0$ such that*

$$P(|\xi_{t_1} - \xi_{t_2}| > \varepsilon) < \gamma,$$

whenever $t_1, t_2 \in [0,T]$ and $|t_1 - t_2| \leq \delta$;

(ii) *it is bounded in probability on $[0,T]$.*

The proof of this theorem is left to the reader as an exercise.

From this point on we will only consider stochastically continuous processes on $[0,\infty)$, without specifying this each time.

To prove that processes with independent increments admit modifications without second-type discontinuities, we need the following lemma.

6. Lemma (Ottaviani's inequality). *Let η_k, $k = 1, ..., n$, be independent random variables, $S_k = \eta_1 + ... + \eta_k$, $a \geq 0$, $0 \leq \alpha < 1$, and*

$$P\{|S_n - S_k| \geq a\} \leq \alpha \quad \forall k.$$

Then for all $c \geq 0$

$$P\{\max_{k \leq n} |S_k| \geq a + c\} \leq \frac{1}{1-\alpha} P\{|S_n| \geq c\}. \tag{1}$$

Proof. The probability on the left in (1) equals

$$\sum_{k=1}^{n} P\{|S_i| < a + c, i < k, |S_k| \geq a + c\}$$

$$\leq \frac{1}{1-\alpha} \sum_{k=1}^{n} P\{|S_i| < a + c, i < k, |S_k| \geq a + c, |S_n - S_k| < a\}$$

$$\leq \frac{1}{1-\alpha} \sum_{k=1}^{n} P\{|S_i| < a+c, i < k, |S_k| \geq a+c, |S_n| \geq c\} \leq \frac{1}{1-\alpha} P\{|S_n| \geq c\}.$$

The lemma is proved.

7. Theorem. *Let ξ_t be a process with independent increments on $[0, \infty)$, $T \in [0, \infty)$, and let ρ be the set of all rational points on $[0, T]$. Then*

$$P\{\sup_{r \in \rho} |\xi_r| < \infty\} = 1.$$

Proof. Obviously it suffices to prove that for some $h > 0$ and all $t \in [0, T]$ we have

$$P\{\sup_{r \in [t, t+h] \cap \rho} |\xi_r| < \infty\} = 1. \tag{2}$$

Take $h > 0$ so that $P\{|\xi_u - \xi_{u+s}| \geq 1\} \leq 1/2$ for all s, u such that $0 \leq s \leq h$ and $s + u \leq T$. Such a choice is possible owing to the uniform stochastic continuity of ξ_t on $[0, T]$. Fix $t \in [0, T]$ and let

$$r_1, ..., r_n \in [t, t + h] \cap \rho, \quad r_1 \leq ... \leq r_n.$$

Observe that $\xi_{r_k} = \xi_{r_1} + (\xi_{r_2} - \xi_{r_1}) + ... + (\xi_{r_k} - \xi_{r_{k-1}})$, where the summands are independent. In addition, $P\{|\xi_{r_n} - \xi_{r_k}| \geq 1\} \leq 1/2$. Hence by Lemma 6

$$P\{\sup_{k\le n}|\xi_{r_k}|\ge 1+c\}\le 2\sup_{t\in[0,T]}P\{|\xi_t|\ge c\}. \tag{3}$$

The last inequality is true for any arrangement of the points $r_k\in[t,t+h]\cap\rho$ which may not be necessarily ordered increasingly. Therefore, now we can think of the set $\{r_1,r_2,...\}$ as being the whole $\rho\cap[t,t+h]$. Then, passing to the limit in (3) as $n\to\infty$ and noticing that

$$\sup\{|\xi_{r_k}|:k=1,2,...\}\uparrow\sup\{|\xi_r|:r\in\rho\cap[t,t+h]\},$$

we find that

$$P\{\sup_{r\in[t,t+h]\cap\rho}|\xi_r|>1+c\}\le 2\sup_{t\in[0,T]}P\{|\xi_t|\ge c\}.$$

Finally, by letting $c\to\infty$ and using the uniform boundedness of ξ_r in probability, we come to (2). The theorem is proved.

Define $D[0,\infty)$ to be the set of all complex-valued right-continuous functions on $[0,\infty)$ which have finite left limits at each point $t\in(0,\infty)$. Similarly one defines $D[0,T]$. We say that a function $x_\cdot$ is a *cadlag* function on $[0,T]$ if $x_\cdot\in D[0,T]$, and just cadlag if $x_\cdot\in D[0,\infty)$.

8. Exercise*. Prove that if $x^n_\cdot\in D[0,\infty)$, $n=1,2,...$, and the x^n_t converge to x_t as $n\to\infty$ uniformly on each finite time interval, then $x_\cdot\in D[0,\infty)$.

9. Lemma. *Let $\rho=\{r_1,r_2,...\}$ be the set of all rational points on $[0,1]$, x_t a real-valued (nonrandom) function given on ρ. For $a<b$ define $\beta_n(x_\cdot,a,b)$ to be the number of upcrossings of the interval (a,b) by the function x_t restricted to the set $r_1,r_2,...,r_n$. Assume that*

$$\lim_{n\to\infty}\beta_n(x_\cdot,a,b)<\infty$$

for any rational a and b. Then the function

$$\tilde x_t:=\lim_{\rho\ni r\downarrow t}x_r$$

is well defined for any $t\in[0,1)$, is right continuous on $[0,T)$, and has (perhaps infinite) left limits on $(0,T]$.

This lemma is set as an exercise on properties of $\overline{\lim}$ and $\underline{\lim}$.

10. Lemma. *Let $\psi(t,\lambda)$ be a complex-valued function defined for $\lambda \in \mathbb{R}$ and $t \in [0,1]$. Assume that $\psi(t,\lambda)$ is continuous in t and never takes the zero value. Let ξ_t be a stochastically continuous process such that*

(i) $\sup_{r\in\rho} |\xi_r| < \infty$ *(a.s.);*

(ii) $\lim_{n\to\infty} E\beta_n(\eta^i_\cdot(\lambda), a, b) < \infty$ *for any* $-\infty < a < b < \infty$, $\lambda \in \mathbb{R}$, $i = 1, 2$, *where*

$$\eta^1_t(\lambda) = \mathrm{Re}\,[\psi(t,\lambda)e^{i\lambda\xi_t}], \quad \eta^2_t(\lambda) = \mathrm{Im}\,[\psi(t,\lambda)e^{i\lambda\xi_t}].$$

Then the process ξ_t admits a modification, all trajectories of which belong to $D[0,1]$.

Proof. Denote $\eta_t(\lambda) = \psi(t,\lambda)e^{i\lambda\xi_t}$ and

$$\Omega' = \bigcap_{m=1}^{\infty} \bigcap_{\substack{a<b \\ a,b \text{ rational}}} \{\lim_{n\to\infty} \beta_n(\eta^i_\cdot(\tfrac{1}{m}), a, b) < \infty, i = 1, 2\} \cap \{\sup_{r\in\rho} |\xi_r| < \infty\}.$$

Obviously, $P(\Omega') = 1$. For $\omega \in \Omega'$ Lemma 9 allows us to let

$$\tilde{\eta}_t(\tfrac{1}{m}) = \lim_{\rho\ni r\downarrow t} \eta_r(\tfrac{1}{m}), \quad t < 1, \quad \tilde{\eta}_1(\tfrac{1}{m}) = \eta_1(\tfrac{1}{m}).$$

For $\omega \notin \Omega'$ let $\tilde{\eta}_t(\frac{1}{m}) \equiv 0$. Observe that, since ψ is continuous in t and $P(\Omega') = 1$ and ξ_t is stochastically continuous, we have that

$$\begin{aligned} &\tilde{\eta}_t(\tfrac{1}{m}) = P\text{-}\lim_{\rho\ni r\downarrow t} \eta_r(\tfrac{1}{m}) = \eta_t(\tfrac{1}{m}) \quad \text{(a.s.)} \quad \forall t < 1, \\ &\tilde{\eta}_1(\tfrac{1}{m}) - \eta_1(\tfrac{1}{m}). \end{aligned} \tag{4}$$

Furthermore, $|\tilde{\eta}_t(\frac{1}{m})\psi^{-1}(t,\frac{1}{m})| \le 1$ for all ω and t.

Now define $\mu = \mu(\omega) = [\sup_{r\in\rho} |\xi_r|] + 1$ and

$$\tilde{\xi}_t = \mu \arcsin \mathrm{Im}\,\tilde{\eta}_t(1/\mu)\psi^{-1}(t, 1/\mu) I_{\Omega'}.$$

By Lemma 9, $\tilde{\eta}_\cdot(\frac{1}{m}) \in D[0,1]$ for any ω. Hence, $\tilde{\xi}_t \in D[0,1]$ for any ω.

It only remains to prove that $P\{\tilde{\xi}_t = \xi_t\} = 1$ for any $t \in [0,1]$. For $t \in \rho$ we have this equality from (4) and from the formula

$$\xi_t = \mu \arcsin \mathrm{Im}\,\eta_t(1/\mu)\psi^{-1}(t, 1/\mu),$$

which holds for $\omega \in \Omega'$. For other t, owing to the stochastic continuity of ξ_t and the right continuity of $\tilde{\xi}_t$, we have

$$\xi_t = P\text{-}\lim_{\rho\ni r\downarrow t}\xi_r = P\text{-}\lim_{\rho\ni r\downarrow t}\tilde{\xi}_r = \tilde{\xi}_t$$

(a.s.). The lemma is proved.

11. Theorem. *Stochastically continuous processes with independent increments admit modifications which are right continuous and have finite left limits for any ω.*

Proof. Let ξ_t be a process in question. It suffices to construct a modification with the described properties on each interval $[n, n+1]$, $n = 0, 1, 2, ...$ The reader can easily combine these modifications to get what we want on $[0,\infty)$. We will confine ourselves to the case $n = 0$. Let ρ be the set of all rational points on $[0, 1]$, and let

$$\varphi(t,\lambda) = Ee^{i\xi_t\lambda}, \quad \varphi(t_1, t_2, \lambda) = Ee^{i\lambda(\xi_{t_2}-\xi_{t_1})}.$$

Since the process ξ_t is stochastically continuous, the function $\varphi(t_1, t_2, \lambda)$ is continuous in $(t_1, t_2) \in [0, 1] \times [0, 1]$ for any λ. Therefore, this function is uniformly continuous on $[0, 1] \times [0, 1]$, and, because $\varphi(t, t, \lambda) = 1$, there exists $\delta(\lambda) > 0$ such that $|\varphi(t_1, t_2, \lambda)| \geq 1/2$ whenever $|t_1 - t_2| < \delta(\lambda)$ and $t_1, t_2 \in [0, 1]$. Furthermore, for any $t \in [0, 1]$ and $\lambda \in \mathbb{R}$ one can find $n \geq 1$ and $0 = t_1 \leq t_2 \leq ... \leq t_n = t$ such that $|t_k - t_{k-1}| < \delta(\lambda)$. Then, using the independence of increments, we find that

$$\varphi(t,\lambda) = \varphi(t_1, t_2, \lambda) \cdot ... \cdot \varphi(t_{n-1}, t_n, \lambda),$$

which implies that $\varphi(t, \lambda) \neq 0$. In addition, $\varphi(t, \lambda)$ is continuous in t.

For fixed λ consider the process

$$\eta_t = \eta_t(\lambda) = \varphi^{-1}(t,\lambda)e^{i\lambda\xi_t}.$$

Let $s_1, s_2, ..., s_n$ be rational numbers in $[0, 1]$ such that $s_1 \leq ... \leq s_n$. Define

$$\mathcal{F}_k = \sigma\{\xi_{s_1}, \xi_{s_2} - \xi_{s_1}, ..., \xi_{s_k} - \xi_{s_{k-1}}\}.$$

Notice that $(\operatorname{Re}\eta_{s_k}, \mathcal{F}_k)$ and $(\operatorname{Im}\eta_{s_k}, \mathcal{F}_k)$ are martingales. Indeed, by virtue of the independence of $\xi_{s_{k+1}} - \xi_{s_k}$ and $\mathcal{F}_k$, we have

$$E\{\operatorname{Re}\eta_{s_{k+1}}|\mathcal{F}_k\} = \operatorname{Re}E\{e^{i\lambda\xi_{s_k}}\varphi^{-1}(s_{k+1},\lambda)e^{i\lambda(\xi_{s_{k+1}}-\xi_{s_k})}|\mathcal{F}_k\}$$

$$= \operatorname{Re}e^{i\lambda\xi_{s_k}}\varphi^{-1}(s_{k+1},\lambda)\varphi(s_k, s_{k+1},\lambda) = \operatorname{Re}\eta_{s_k} \quad \text{(a.s.)}.$$

Hence by Doob's upcrossing theorem, if $r_i \in \rho$, $\{r_1, ..., r_n\} = \{s_1, ..., s_n\}$, and $0 \leq s_1 \leq ... \leq s_n$, then

$$E\beta_n(\operatorname{Re}\eta_\cdot, a, b) \leq (E|\operatorname{Re}\eta_{s_n}| + |a|)/(b-a)$$

$$\leq (\sup_{t\in[0,1]} \varphi^{-1}(t,\lambda) + |a|)/(b-a) < \infty,$$

$$\sup_n E\beta_n(\operatorname{Im}\eta_\cdot, a, b) < \infty.$$

It only remains to apply Lemma 10. The theorem is proved.

12. Exercise* (cf. Exercise 3). Take the stable process τ_a, $a \geq 0$, from Theorem 2.6.1. Observe that τ_a increases in a and prove that its cadlag modification, the existence of which is asserted in Theorem 11, is given by τ_{a+}, $a \geq 0$.

2. Lévy-Khinchin theorem

In this section we prove a remarkable Lévy-Khinchin theorem. It is worth noting that this theorem was originally proved for so-called infinitely divisible laws and not for infinitely divisible processes. As usual we are only dealing with one-dimensional processes (the multidimensional case is treated, for instance, in [**GS**]).

1. Definition. A process ξ_t with independent increments is called *time homogeneous* if, for every $h > 0$, the distribution of $\xi_{t+h} - \xi_t$ is independent of t.

2. Definition. A stochastically continuous time-homogeneous process ξ_t with independent increments is called an *infinitely divisible process*.

3. Theorem (Lévy-Khinchin). *Let ξ_t be an infinitely divisible process on $[0,\infty)$. Then there exist a finite nonnegative measure on $(\mathbb{R}, \mathfrak{B}(\mathbb{R}))$ and a number $b \in \mathbb{R}$ such that, for any $t \in [0,\infty)$ and $\lambda \in \mathbb{R}$, we have*

$$Ee^{i\lambda\xi_t} = \exp\{t\int_{\mathbb{R}} f(\lambda,x)\,\mu(dx) + itb\lambda\}, \tag{1}$$

where

$$f(\lambda,x) = (e^{i\lambda x} - 1 - i\lambda\sin x)\frac{1+x^2}{x^2}, \quad x \neq 0, \quad f(\lambda,0) := -\frac{\lambda^2}{2}.$$

Proof. Denote $\varphi(t,\lambda) = Ee^{i\lambda\xi_t}$. In the proof of Theorem 1.11 we saw that $\varphi(t,\lambda)$ is continuous in t and $\varphi(t,\lambda) \neq 0$. In addition $\varphi(t,\lambda)$ is continuous with respect to the pair (t,λ). Define

$$a(t,\lambda) = \arg\varphi(t,\lambda), \quad l(t,\lambda) = \ln|\varphi(t,\lambda)|.$$

By using the continuity of φ and the fact that $\varphi \neq 0$, one can uniquely define $a(t,\lambda)$ to be continuous in t and in λ and satisfy $a(0,\lambda) = a(t,0) = 0$.

Clearly, $l(t,\lambda)$ is a finite function which is also continuous in t and in λ. Furthermore,

$$\varphi(t,\lambda)=\exp\{l(t,\lambda)+ia(t,\lambda)\}.$$

Next, it follows from the homogeneity and independence of increments of ξ_t that

$$\varphi(t+s,\lambda)=\varphi(t,\lambda)\varphi(s,\lambda).$$

Hence, by definition of a, we get that, for each λ, it satisfies the equation

$$f(t+s)=f(t)+f(s)+2\pi k(s,t),$$

where $k(s,t)$ is a continuous integer-valued function. Since $k(t,0)=0$, in fact, $k\equiv 0$, and a satisfies $f(t+s)=f(t)+f(s)$. The same equation is also valid for l. Any continuous solution of this equation has the form ct, where c is a constant. Thus,

$$a(t,\lambda)=ta(\lambda),\quad l(t,\lambda)=tl(\lambda),$$

where $a(\lambda)=a(1,\lambda)$ and $l(\lambda)=l(1,\lambda)$. By defining $g(\lambda):=l(\lambda)+ia(\lambda)$, we write

$$\varphi(t,\lambda)=e^{tg(\lambda)},$$

where g is a continuous function of λ and $g(0)=0$. We have reduced our problem to finding g.

Observe that

$$g(\lambda)=\lim_{t\downarrow 0}\frac{e^{tg(\lambda)}-1}{t}=\lim_{t\downarrow 0}\frac{\varphi(t,\lambda)-1}{t}. \tag{2}$$

Moreover, from Taylor's expansion of $\exp(tg(\lambda))$ with respect to t one easily sees that the convergence in (2) is uniform on each set of values of λ on which $g(\lambda)$ is bounded. In particular, this is true on each set $[-h,h]$ with $0\le h<\infty$.

By taking t of type $1/n$ and denoting F_t the distribution of ξ_t, we conclude that

$$n\int_{\mathbb{R}}(e^{i\lambda x}-1)\,F_{1/n}(dx)\to g(\lambda) \tag{3}$$

as $n\to\infty$ uniformly in λ on any finite interval. Integrate this against $d\lambda$ to get

$$\lim_{n\to\infty} n \int_{\mathbb{R}} \left(1 - \frac{\sin xh}{xh}\right) F_{1/n}(dx) = -\frac{1}{2h} \int_{-h}^{h} g(\lambda)\, d\lambda. \qquad (4)$$

Notice that the right-hand side of (4) can be made arbitrarily small by choosing h small, since g is continuous and vanishes at zero. Furthermore, as is easy to see, $1 - \sin xh/xh \geq 1/2$ for $|xh| \geq 2$. It follows that, for any $\varepsilon > 0$, there exists $h > 0$ such that

$$\overline{\lim_{n\to\infty}}\, (n/2) \int_{|x|\geq 2/h} F_{1/n}(dx) \leq \varepsilon.$$

In turn, it follows that, for all large n,

$$n \int_{|x|\geq 2/h} F_{1/n}(dx) \leq 4\varepsilon. \qquad (5)$$

By reducing h one can accommodate any finite set of values of n and find an h such that (5) holds for all $n \geq 1$ rather than only for large ones.

To derive yet another consequence of (4), notice that there exists a constant $\gamma > 0$ such that

$$1 - \frac{\sin x}{x} \geq \gamma \frac{x^2}{1+x^2} \qquad \forall x \in \mathbb{R}.$$

Therefore, from (4) with $h = 1$, we obtain that there exists a finite constant c such that for all n

$$n \int_{\mathbb{R}} \frac{x^2}{1+x^2} F_{1/n}(dx) \leq c. \qquad (6)$$

Finally, upon introducing measures μ_n by the formula

$$\mu_n(dx) = n \frac{x^2}{1+x^2} F_{1/n}(dx),$$

and noticing that $\mu_n \leq nF_{1/n}$, from (5) and (6), we see that the family $\{\mu_n, n = 1, 2, ...\}$ is weakly compact. Therefore, there exist a subsequence $n' \to \infty$ and a finite measure μ such that

$$\int_{\mathbb{R}} f(x)\, \mu_{n'}(dx) \to \int_{\mathbb{R}} f(x)\, \mu(dx)$$

for every bounded and continuous f. As is easy to check $f(\lambda, x)$ is bounded and continuous in x. Hence,

$$g(\lambda) = \lim_{n\to\infty} n \int_{\mathbb{R}} (e^{i\lambda x} - 1)\, F_{1/n}(dx)$$

$$= \lim_{n\to\infty} [\int_{\mathbb{R}} f(\lambda, x)\mu_n(dx) + i\lambda n \int_{\mathbb{R}} \sin x\, F_{1/n}(dx)]$$

$$= \lim_{n'\to\infty} [\int_{\mathbb{R}} f(\lambda, x)\mu_{n'}(dx) + i\lambda n' \int_{\mathbb{R}} \sin x\, F_{1/n'}(dx)]$$

$$= \int_{\mathbb{R}} f(\lambda, x)\mu(dx) + i\lambda b,$$

where

$$b := \lim_{n'\to\infty} n' \int_{\mathbb{R}} \sin x\, F_{1/n'}(dx),$$

and the existence and finiteness of this limit follows from above computations in which all limits exist and are finite. The theorem is proved.

Formula (1) is called *Khinchin's formula*. The following *Lévy's formula* sheds more light on the structure of the process x_t:

$$\varphi(t, \lambda) = \exp t\{\int_{\mathbb{R}} (e^{i\lambda x} - 1 - i\lambda \sin x)\, \Lambda(dx) + ib\lambda - \sigma^2\lambda^2/2\},$$

where Λ is called *the Lévy measure of* ξ_t. This is a nonnegative, generally speaking, infinite measure on $\mathfrak{B}(\mathbb{R})$ such that

$$\int_{\mathbb{R}} \frac{x^2}{1+x^2}\, \Lambda(dx) < \infty, \quad \Lambda(\{0\}) = 0. \tag{7}$$

Any such measure is called *a Lévy measure*. One obtains one formula from the other by introducing the following relations between μ and the pair (Λ, σ^2):

$$\mu(\{0\}) = \sigma^2, \quad \Lambda(\Gamma) = \int_{\Gamma\setminus\{0\}} \frac{1+x^2}{x^2}\, \mu(dx).$$

4. Exercise*. Prove that if one introduces (Λ, σ^2) by the above formulas, then one gets Lévy's formula from Khinchin's formula, and, in addition, Λ satisfies (7).

5. Exercise*. Let a measure Λ satisfy (7). Define

$$\mu(\Gamma) = \int_\Gamma \frac{x^2}{1+x^2}\,\Lambda(dx) + I_\Gamma(0)\sigma^2.$$

Show that μ is a finite measure for which Lévy's formula transforms into Khinchin's formula.

6. Theorem (uniqueness). *There can exist only one finite measure μ and one number b for which $\varphi(t,\lambda)$ is representable by Khinchin's formula. There can exist only one measure Λ satisfying (7) and unique numbers b and σ^2 for which $\varphi(t,\lambda)$ is representable by Lévy's formula.*

Proof. Exercises 4 and 5 show that we may concentrate only on the first part of the theorem. The exponent in Khinchin's formula is continuous in λ and vanishes at $\lambda = 0$. Therefore it is uniquely determined by $\varphi(t,\lambda)$, and we only need prove that μ and b are uniquely determined by the function

$$g(\lambda) := \int_{\mathbb{R}} f(\lambda, x)\,\mu(dx) + ib\lambda.$$

Clearly, it suffices only to show that μ is uniquely determined by g.

For $h > 0$, we have

$$g(\lambda) - \frac{g(\lambda+h)+g(\lambda-h)}{2} = \int_{\mathbb{R}} e^{i\lambda x}\,\frac{1-\cos xh}{x^2}\,(1+x^2)\,\mu(dx) \tag{8}$$

with the agreement that $(1-\cos xh)/x^2 = h^2/2$ if $x = 0$. Define a new measure

$$\nu_h(\Gamma) = \int_\Gamma \rho(x,h)\,\mu(dx), \quad \rho(x,h) = \frac{1-\cos xh}{x^2}\,(1+x^2)$$

and use

$$\int_{\mathbb{R}} f(x)\,\nu_h(dx) = \int_{\mathbb{R}} f(x)\rho(x,h)\,\mu(dx)$$

for all bounded Borel f. Then we see from (8) that the characteristic function of ν_h is uniquely determined by g. Therefore, ν_h is uniquely determined by g for any $h > 0$.

Now let Γ be a bounded Borel set and h be such that $\Gamma \subset [-1/h, 1/h]$. Take $f(x) = \rho^{-1}(x,h)$ for $x \in \Gamma$ and $f(x) = 0$ elsewhere. By the way, observe that f is a bounded Borel function. For this f

$$\int_{\mathbb{R}} f(x)\,\nu_h(dx) = \int_{\mathbb{R}} f(x)\rho(x,h)\,\mu(dx) = \mu(\Gamma),$$

where the left-hand side is uniquely determined by g. The theorem is proved.

7. Corollary. *Define*

$$\mu_t(dx) = \frac{x^2}{t(1+x^2)}\,F_t(dx), \quad b_t = \frac{1}{t}\int_{\mathbb{R}} \sin x\, F_t(dx).$$

Then $\mu_t \to \mu$ *weakly and* $b_t \to b$ *as* $t \downarrow 0$.

Indeed, similarly to (3) we have

$$\frac{1}{t}\int_{\mathbb{R}} (e^{i\lambda x} - 1)\, F_t(dx) \to g(\lambda),$$

which as in the proof of the Lévy-Khinchin theorem shows that the family $\{\mu_t; t \le 1\}$ is weakly compact. Next, if $\mu_{t_{n'}} \xrightarrow{w} \nu$, then, again as in the proof of the Lévy-Khinchin theorem, $b_{t_{n'}}$ converges, and if we denote its limit by c, then Khinchin's formula holds with $\mu = \nu$ and $b = c$. Finally, the uniqueness implies that all weak limit points of μ_t, $t \downarrow 0$, coincide with μ and hence (cf. Exercise 1.2.10) $\mu(t) \xrightarrow{w} \mu$ as $t \downarrow 0$. This obviously implies that b_t also converges and its limit is b.

8. Corollary. *In Lévy's formula*

$$\sigma^2 = \lim_{n\to\infty} \lim_{t\downarrow 0} \frac{1}{t}\, E\xi_t^2 I_{|\xi_t| \le \varepsilon_n},$$

where ε_n *is a sequence such that* $\varepsilon_n > 0$ *and* $\varepsilon_n \downarrow 0$. *Moreover,* F_t/t *converges weakly on* $\mathbb{R} \setminus \{0\}$ *as* $t \downarrow 0$ *to* Λ, *that is,*

$$\lim_{t\downarrow 0} \frac{1}{t}\int_{\mathbb{R}} f(x)\, F_t(dx) = \lim_{t\downarrow 0} \frac{1}{t}\, Ef(\xi_t) = \int_{\mathbb{R}} f(x)\, \Lambda(dx) \tag{9}$$

for each bounded continuous function f *which vanishes in a neighborhood of* 0.

Proof. By the definition of Λ and Corollary 7, for each bounded continuous function f which vanishes in a neighborhood of 0, we have

$$\int_{\mathbb{R}} f(x)\, \Lambda(dx) = \int_{\mathbb{R}} f(x)\, \frac{1+x^2}{x^2}\, \mu(dx)$$

$$= \lim_{t\downarrow 0} \int_{\mathbb{R}} f(x)\, \frac{1+x^2}{x^2}\, \mu_t(dx) = \lim_{t\downarrow 0} \frac{1}{t}\int_{\mathbb{R}} f(x)\, F_t(dx).$$

This proves (9).

Let us prove the first assertion. By the dominated convergence theorem, for every sequence of nonnegative $\varepsilon_n \to 0$ we have

$$\sigma^2 = \mu(\{0\}) = \int_{\mathbb{R}} I_{[0,0]}(x)\,\mu(dx)$$

$$= \int_{\mathbb{R}} I_{[0,0]}(x)(1+x^2)\,\mu(dx) = \lim_{n\to\infty} \int_{\mathbb{R}} I_{[-\varepsilon_n,\varepsilon_n]}(x)(1+x^2)\,\mu(dx).$$

By Theorem 1.2.11 (v), if $\mu(\{\varepsilon_n\}) = \mu(\{-\varepsilon_n\}) = 0$, then

$$\int_{\mathbb{R}} I_{[-\varepsilon_n,\varepsilon_n]}(x)(1+x^2)\,\mu(dx)$$

$$= \lim_{t\downarrow 0} \frac{1}{t} \int_{\mathbb{R}} I_{[-\varepsilon_n,\varepsilon_n]}(x)x^2\,F_t(dx) = \lim_{t\downarrow 0} \frac{1}{t}\,E\xi_t^2 I_{|\xi_t|\le\varepsilon_n}.$$

It only remains to notice that the set of x such that $\mu(\{x\}) > 0$ is countable, so that there exists a sequence ε_n such that $\varepsilon_n \downarrow 0$ and $\mu(\{\varepsilon_n\}) = \mu(\{-\varepsilon_n\}) = 0$. The corollary is proved.

9. Exercise. Prove that, if $\xi_t \ge 0$ for all $t \ge 0$ and ω, then $\Lambda((-\infty,0]) = 0$. One can say more in that case, as we will see in Exercise 3.15.

We know that the Wiener process has independent increments, and also that it is homogeneous and stochastically continuous (even just continuous). In Lévy's formula, to get $E\exp(i\lambda w_t)$ one takes $\Lambda = 0$, $b = 0$, and $\sigma = 1$.

If in Lévy's formula we take $\sigma = 0$, $\Lambda(\Gamma) = I_\Gamma(1)\mu$, and $b = \mu\sin 1$, where μ is a nonnegative number, then the corresponding process is called *the Poisson process with parameter* μ.

If $\sigma = b = 0$ and $\Lambda(dx) = ax^{-2}\,dx$ with a constant $a > 0$, the corresponding process is called *the Cauchy process*.

Clearly, for the Poisson process π_t with parameter μ we have

$$Ee^{i\lambda\pi_t} = e^{t\mu(e^{i\lambda}-1)},$$

so that π_t has Poisson distribution with parameter $t\mu$. In particular,

$$E|\pi_{t+h} - \pi_t| = E\pi_h = h\mu$$

for $t, h \ge 0$. The values of π_t are integers and π_t is not identically constant (the expectation grows). Therefore π_t does not have continuous modification, which shows, in particular, that the requirement $\beta > 0$ in Theorem 1.4.8 is essential. For $\mu = 1$ we come to the Poisson process introduced in Exercise 2.3.8.

10. Exercise. Prove that for the Cauchy process we have $\varphi(t, \lambda) = \exp(-ct|\lambda|)$, with a constant $c > 0$.

11. Exercise*. Prove that the Lévy measure of the process τ_{a+}, $a \geq 0$ (see Theorem 2.6.1, and Exercise 1.12) is concentrated on the positive half line and is given by $I_{x>0}(2\pi)^{-1/2}x^{-3/2}\,dx$. This result will be used in Sec. 6.

You may also like to show that

$$\varphi(t, \lambda) = \exp(-t|\lambda|^{1/2}(a - ib \operatorname{sign} \lambda)),$$

where

$$a = (2\pi)^{-1/2} \int_0^\infty x^{-3/2}(1 - \cos x)\,dx, \quad b = (2\pi)^{-1/2} \int_0^\infty x^{-3/2} \sin x\,dx,$$

and, furthermore, that $a = b = 1$.

12. Exercise. Prove that if in Lévy's formula we have $\Lambda = 0$ and $\sigma = 0$, then $\xi_t = bt$ (a.s.) for all t, where b is a constant.

3. Jump measures and their relation to Lévy measures

Let ξ_t be an infinitely divisible cadlag process on $[0, \infty)$. Define

$$\Delta\xi_t = \xi_t - \xi_{t-}.$$

For any set $\Gamma \subset \mathbb{R}_+ \times \mathbb{R} := [0, \infty) \times \mathbb{R}$ let $p(\Gamma)$ be the number of points $(t, \Delta\xi_t) \in \Gamma$. It may happen that $p(\Gamma) = \infty$. Obviously $p(\Gamma)$ is a σ-additive measure on the family of all subsets of $\mathbb{R}_+ \times \mathbb{R}$. The measure $p(\Gamma)$ is called *the jump measure of* ξ_t.

For $T, \varepsilon \in (0, \infty)$ define

$$R_{T,\varepsilon} = [0, T] \times \{x : |x| \geq \varepsilon\}.$$

1. Remark. Notice that $p(R_{T,\varepsilon}) < \infty$ for any ω, which is to say that on $[0, T]$ there may be only finitely many t such that $|\Delta\xi_t| \geq \varepsilon$. This property follows immediately from the fact that the trajectories of ξ_t do not have discontinuities of the second kind. It is also worth noticing that $p(\Gamma)$ is concentrated at points $(t, \Delta\xi_t)$ and each point of this type receives a unit mass.

We will need yet another measure defined on subsets of $\mathbb{R}$. For any $B \subset \mathbb{R}$ define

$$p_t(B) = p((0, t] \times B).$$

2. Remark. By Remark 1, if B is separated from zero, then $p_t(B)$ is finite. Moreover, let $f(x)$ be a Borel function (perhaps unbounded) vanishing for $|x| < \varepsilon$, where $\varepsilon > 0$. Then, the process

$$\eta_t := \eta_t(f) := \int_{\mathbb{R}} f(x)\, p_t(dx)$$

is well defined and is just equal to the (finite) sum of $f(\Delta\xi_s)$ for all $s \leq t$ such that $|\Delta\xi_s| \geq \varepsilon$.

The structure of η_t is pretty simple. Indeed, fix an ω and let $0 \leq s_1 < ... < s_n < ...$ be all s for which $|\Delta\xi_s| \geq \varepsilon$ (if there are only $m < \infty$ such s, we let $s_n = \infty$ for $n \geq m+1$). Then, of course, $s_n \to \infty$ as $n \to \infty$. Also $s_1 > 0$, because ξ_t is right continuous and $\xi_0 = 0$. With this notation

$$\eta_t = \sum_{s_n \leq t} f(\Delta\xi_{s_n}). \tag{1}$$

We see that η_t starts from zero, is constant on each interval $[s_{n-1}, s_n)$, $n = 1, 2, ...$ (with $s_0 := 0$), and

$$\Delta\eta_{s_n} = f(\Delta\xi_{s_n}). \tag{2}$$

3. Lemma. *Let $f(x)$ be a function as in Remark 2. Assume that f is continuous. Let $0 \leq t < \infty$ and t_i^n be such that*

$$s = t_1^n < ... < t_{k(n)+1}^n = t, \qquad \max_{j=1,...,k(n)} (t_{j+1}^n - t_j^n) \to 0$$

as $n \to \infty$. Then for any ω

$$\eta_t(f) - \eta_s(f) = \int_{\mathbb{R}_+ \times \mathbb{R}} I_{(s,t]}(u) f(x)\, p(dudx) = \lim_{n\to\infty} \sum_{j=1}^{k(n)} f(\xi_{t_{j+1}^n} - \xi_{t_j^n}). \tag{3}$$

Proof. We have noticed above that the set of all $u \in (s,t]$ for which $|\Delta\xi_u| \geq \varepsilon$ is finite. Let $\{u_1, ..., u_N\}$ be this set. Single out those intervals $(t_j^h, t_{j+1}^n]$ which contain at least one of the u_i's. For large n we will have exactly N such intervals. First we prove that, for large n,

$$|\xi_{t_{j+1}^n} - \xi_{t_j^n}| < \varepsilon, \quad f(\xi_{t_{j+1}^n} - \xi_{t_j^n}) = 0$$

if the interval $(t_j^n, t_{j+1}^n]$ does not contain any of the u_i's. Indeed, if this were not true, then one could find a sequence s_k, t_k such that $|\xi_{t_k} - \xi_{s_k}| \geq \varepsilon$, $s_k, t_k \in (s,t]$, $s_k < t_k$, $t_k - s_k \to 0$, and on $(s_k, t_k]$ there are no points u_i.

Without loss of generality, we may assume that $s_k, t_k \to u \in (s,t]$ (actually, one can obviously assume that $u \in [s,t]$, but since the trajectories are right continuous, $\xi_{s_k}, \xi_{t_k} \to \xi_s$ if $s_k, t_k \to s$, so that $u \neq s$).

Furthermore, there are infinitely many s_k's either to the right or to the left of u. Therefore, using subsequences if needed, we may assume that the sequence s_k is monotone and then that t_k is monotone as well. Then, since ξ_t has finite right and left limits, we have that $s_k \uparrow u$, $s_k < u$, and $t_k \downarrow u$, which implies that $|\Delta\xi_u| \geq \varepsilon$. But then we would have a point $u \in \{u_1, ..., u_N\}$ which belongs to $(s_k, t_k]$ for all k (after passing to subsequences). This is a contradiction, which proves that for all large n the sum on the right in (3) contains at most N nonzero terms. These terms correspond to the intervals $(t_j, t_{j+1}]$ containing u_i's, and they converge to $f(\Delta\xi_{u_i})$.

It only remains to observe that the first equality in (3) is obvious and, by Remark 2,

$$\int_{\mathbb{R}_+ \times \mathbb{R}} I_{(s,t]}(u) f(x)\, p(dudx) = \sum_{i=1}^{N} f(\Delta\xi_{u_i}).$$

The lemma is proved.

4. Definition. For $0 \leq s < t < \infty$ define $\mathcal{F}^\xi_{s,t}$ as the completion of the σ-field generated by $\xi_r - \xi_s$, $r \in [s,t]$. Also set $\mathcal{F}^\xi_t = \mathcal{F}^\xi_{0,t}$.

5. Remark. Since the increments of ξ_t are independent, the σ-fields $\mathcal{F}^\xi_{0,t_1}$, $\mathcal{F}^\xi_{t_1,t_2}, ..., \mathcal{F}^\xi_{t_{n-1},t_n}$ are independent for any $0 < t_1 < ... < t_n$.

Next remember Definition 2.5.10.

6. Definition. Random processes $\eta^1_t, ..., \eta^n_t$ defined for $t \geq 0$ are called *independent* if for any $t_1, ..., t_k \geq 0$ the vectors $(\eta^1_{t_1}, ..., \eta^1_{t_k}), ..., (\eta^n_{t_1}, ..., \eta^n_{t_k})$ are independent.

7. Lemma. *Let ζ_t be an $\mathbb{R}^d$-valued process starting from zero and such that $\zeta_t - \zeta_s$ is $\mathcal{F}^\xi_{s,t}$-measurable whenever $0 \leq s < t < \infty$. Also assume that, for all $0 \leq s < t < \infty$, the random variables $\zeta^1_t - \zeta^1_s, ..., \zeta^d_t - \zeta^d_s$ are independent. Then the process ζ_t has independent increments and the processes $\zeta^1_t, ..., \zeta^d_t$ are independent.*

Proof. That ζ_t has independent increments follows from Remark 5. To prove that the vectors

$$(\zeta^1_{t_1}, ..., \zeta^1_{t_n}), ..., (\zeta^d_{t_1}, ..., \zeta^d_{t_n}) \tag{4}$$

are independent if $0 = t_0 < t_1 < ..., < t_n$, it suffices to prove that

$$(\zeta^1_{t_1} - \zeta^1_{t_0}, \zeta^1_{t_2} - \zeta^1_{t_1}, ..., \zeta^1_{t_n} - \zeta^1_{t_{n-1}}), ..., (\zeta^d_{t_1} - \zeta^d_{t_0}, \zeta^d_{t_2} - \zeta^d_{t_1}, ..., \zeta^d_{t_n} - \zeta^1_{t_{n-1}}) \tag{5}$$

are independent. Indeed, the vectors in (4) can be obtained after applying a linear transformation to the vectors in (5). Now take $\lambda^k_j \in \mathbb{R}$ for $k = 1, ..., d$ and $j = 1, .., n$, and write

$$E\exp\big(i\sum_{k,j}\lambda^k_j(\zeta^k_{t_j} - \zeta^k_{t_{j-1}})\big)$$

$$= E\exp\big(i\sum_{j\le n-1,k}\lambda^k_j(\zeta^k_{t_j} - \zeta^k_{t_{j-1}})\big)E\{\exp\big(i\sum_{k}\lambda^k_n(\zeta^k_{t_n} - \zeta^k_{t_{n-1}})\big)|\mathcal{F}^\xi_{0,t_{n-1}}\}$$

$$= E\exp\big(i\sum_{j\le n-1,k}\lambda^k_j(\zeta^k_{t_j} - \zeta^k_{t_{j-1}})\big)E\exp\big(i\sum_{k}\lambda^k_n(\zeta^k_{t_n} - \zeta^k_{t_{n-1}})\big)$$

$$= E\exp\big(i\sum_{j\le n-1,k}\lambda^k_j(\zeta^k_{t_j} - \zeta^k_{t_{j-1}})\big)\prod_k E\exp\big(i\lambda^k_n(\zeta^k_{t_n} - \zeta^k_{t_{n-1}})\big).$$

An obvious induction allows us to represent the characteristic function of the family $\{\zeta^k_{t_j} - \zeta^k_{t_{j-1}}, k = 1, ..., d, j = 1, ..., n\}$ as the product of the characteristic functions of its members, thus proving the independence of all $\zeta^k_{t_j} - \zeta^k_{t_{j-1}}$ and, in particular, of the vectors (5). The lemma is proved.

8. Lemma. *Let f be as in Remark 2 and let f be continuous. Take $\alpha \in \mathbb{R}$ and denote $\zeta_t = \eta_t(f) + \alpha\xi_t$. Then*

(i) *for every $0 \le s < t < \infty$, the random variable $\zeta_t - \zeta_s$ is $\mathcal{F}^\xi_{s,t}$-measurable;*

(ii) *the process ζ_t is an infinitely divisible cadlag process and*

$$Ee^{i\zeta_t} = \exp t\{\int_{\mathbb{R}}(e^{i(f(x)+\alpha x)} - 1 - i\alpha\sin x)\,\Lambda(dx) + i\alpha b - \alpha^2\sigma^2/2\}. \tag{6}$$

Proof. Assertion (i) is a trivial consequence of (3). In addition, Remark 5 shows that ζ_t has independent increments.

(ii) The homogeneity of ζ_t follows immediately from (3) and the similar property of ξ_t. Furthermore, Remark 2 shows that ζ_t is cadlag. From the homogeneity and right continuity of ζ_t we get

$$\lim_{s\uparrow t} Ee^{i\lambda(\zeta_t - \zeta_s)} = \lim_{s\uparrow t} Ee^{i\lambda\zeta_{t-s}} = Ee^{i\lambda\zeta_0} = 1, \quad t > 0.$$

Similar equations hold for $s \downarrow t$ with $t \geq 0$. Therefore, $\zeta_s \xrightarrow{P} \zeta_t$ as $s \to t$, and ζ_t is stochastically continuous.

To prove (6), take Khinchin's measure μ and take μ_t and b_t from Corollary 2.7. Also observe that

$$\lim_{n\to\infty} a_n^n = \lim_{n\to\infty} e^{n \log a_n} = \lim_{n\to\infty} e^{n(a_n-1)}$$

provided $a_n \to 1$ and one of the limits exists. Then we have

$$Ee^{i(\eta_t+\alpha\xi_t)} = \lim_{n\to\infty} \left(Ee^{i(f(\xi_{t/n})+\alpha\xi_{t/n})}\right)^n$$

$$= \lim_{n\to\infty} \exp n \int_{\mathbb{R}} (e^{i(f(x)+\alpha x)} - 1)\, F_{t/n}(dx),$$

with

$$\lim_{n\to\infty} n \int_{\mathbb{R}} (e^{i(f(x)+\alpha x)} - 1)\, F_{t/n}(dx)$$

$$= \lim_{n\to\infty} t \int_{\mathbb{R}} (e^{i(f(x)+\alpha x)} - 1 - i\alpha \sin x)(1+x^2)/x^2\, \mu_{t/n}(dx) + i\alpha t b$$

$$= t \int_{\mathbb{R}} (e^{i(f(x)+\alpha x)} - 1 - i\alpha \sin x)(1+x^2)/x^2\, \mu(dx) + i\alpha t b.$$

Now to get (6) one only has to refer to Exercise 2.4. The lemma is proved.

9. Theorem. (i) *For $ab > 0$ the process $p_t(a,b]$ is a Poisson process with parameter $\Lambda((a,b])$, and, in particular,*

$$Ep_t(a,b] = t\Lambda((a,b]); \tag{7}$$

(ii) *if $a_m < b_m$, $a_m b_m > 0$, $m = 1, ..., n$, and the intervals $(a_m, b_m]$ are pairwise disjoint, then the processes $p_t(a_1,b_1], ..., p_t(a_n,b_n]$ are independent.*

Proof. To prove (i), take a sequence of bounded continuous functions $f_k(x)$ such that $f_k(x) \to \lambda I_{(a,b]}(x)$ as $k \to \infty$ and $f_k(x) = 0$ for $|x| < \varepsilon := (|a| \wedge |b|)/2$. Then, for each ω,

$$\int_{\mathbb{R}} f_k(x)\, p_t(dx) \to \lambda p_t(a,b]. \tag{8}$$

Moreover, $|\exp\{if_k(x)\} - 1| \leq 2I_{|x|\geq\varepsilon}$ and

$$\int_{\mathbb{R}} I_{|x|\geq\varepsilon}\,\Lambda(dx) \leq \frac{1+\varepsilon^2}{\varepsilon^2}\int_{\mathbb{R}} \frac{x^2}{1+x^2}\,\Lambda(dx) < \infty. \tag{9}$$

Hence, by Lemma 8 and by the dominated convergence theorem,

$$Ee^{i\lambda p_t(a,b]} = \exp t\int_{\mathbb{R}} (e^{i\lambda I_{(a,b]}(x)} - 1)\,\Lambda(dx) = \exp\{t\Lambda((a,b])(e^{i\lambda} - 1)\}.$$

The homogeneity of $p_t(a,b]$ and independence of its increments follow from (8) and Lemma 8. Remark 2 shows that $p_t(a,b]$ is a cadlag process. As in Lemma 8, this leads to the conclusion that $p_t(a,b]$ is stochastically continuous. This proves (i).

(ii) Formula (8) and Lemma 8 imply that $p_t(a,b] - p_s(a,b]$ is $\mathcal{F}^{\xi}_{s,t}$-measurable if $s < t$. By Lemma 7, to prove that the processes $p_t(a_1,b_1]$,..., $p_t(a_n,b_n]$ are independent, it suffices to prove that, for any $s < t$, the random variables

$$p_t(a_1,b_1] - p_s(a_1,b_1], ..., p_t(a_n,b_n] - p_s(a_n,b_n] \tag{10}$$

are independent.

Take $\lambda_1, ..., \lambda_n \in \mathbb{R}$ and define $f(x) = \lambda_m$ for $x \in (a_m,b_m]$ and $f = 0$ outside the union of the $(a_m,b_m]$. Also take a sequence of bounded continuous functions f_n vanishing in a neighborhood of zero such that $f_n(x) \to f(x)$ for all $x \in \mathbb{R}$. Then

$$\eta_t(f_n) - \eta_s(f_n) \to \eta_t(f) - \eta_s(f) = \sum_{m=1}^{n} \lambda_m\{p_t(a_m,b_m] - p_s(a_m,b_m]\}.$$

Hence and from Lemma 8 we get

$$E\exp(i\sum_{m=1}^{n} \lambda_m\{p_t(a_m,b_m] - p_s(a_m,b_m]\}) = \lim_{n\to\infty} Ee^{i(\eta_t(f_n)-\eta_s(f_n))}$$

$$= \lim_{n\to\infty} Ee^{i\eta_{t-s}(f_n)} = \lim_{n\to\infty} \exp\{(t-s)\int_{\mathbb{R}} (e^{if_n(x)} - 1)\,\Lambda(dx)\}$$

$$= \exp\{(t-s)\int_{\mathbb{R}} (e^{if(x)} - 1)\,\Lambda(dx)\} = \prod_{m=1}^{n} \exp\{(t-s)\Lambda((a_m,b_m])(e^{i\lambda_m} - 1)\}.$$

This and assertion (i) prove that the random variables in (5) are independent. The theorem is proved.

10. Corollary. *Let f be a Borel nonnegative function. Then, for each $t \geq 0$,*

$$\int_{\mathbb{R}\setminus\{0\}} f(x)\, p_t(dx)$$

is a random variable and

$$E\int_{\mathbb{R}\setminus\{0\}} f(x)\, p_t(dx) = t\int_{\mathbb{R}} f(x)\, \Lambda(dx). \tag{11}$$

Notice that on the right in (11) we write the integral over $\mathbb{R}$ instead of $\mathbb{R}\setminus\{0\}$ because $\Lambda(\{0\}) = 0$ by definition. To prove the assertion, take $\varepsilon > 0$ and let Σ be the collection of all Borel Γ such that $p_t(\Gamma\setminus(-\varepsilon,\varepsilon))$ is a random variable and

$$\nu_\varepsilon(\Gamma) := Ep_t(\Gamma\setminus(-\varepsilon,\varepsilon)) = t\Lambda_\varepsilon(\Gamma) := t\Lambda(\Gamma\setminus(-\varepsilon,\varepsilon)).$$

It follows from (7) and from the finiteness of $\Lambda(\mathbb{R}\setminus(-\varepsilon,\varepsilon))$ that $\mathbb{R}\in\Sigma$. By adding an obvious argument we conclude that Σ is a λ-system. Furthermore, from Theorem 9 (i) we know that Σ contains $\Pi := \{(a,b] : ab > 0\}$, which is a π-system. Therefore, $\Sigma = \mathfrak{B}(\mathbb{R})$. Now a standard measure-theoretic argument shows that, for every Borel nonnegative f, we have

$$E\int_{\mathbb{R}\setminus(-\varepsilon,\varepsilon)} f(x)\, p_t(dx) = \int_{\mathbb{R}} f(x)\, \nu_\varepsilon(dx)$$

$$= t\int_{\mathbb{R}} f(x)\, \Lambda_\varepsilon(dx) = t\int_{\mathbb{R}\setminus(-\varepsilon,\varepsilon)} f(x)\, \Lambda(dx).$$

It only remains to let $\varepsilon \downarrow 0$ and use the monotone convergence theorem.

11. Corollary. *Every continuous infinitely divisible process has the form $bt + \sigma w_t$, where σ and b are the constants from Lévy's formula and w_t is a Wiener process if $\sigma \neq 0$ and $w_t \equiv 0$ if $\sigma = 0$.*

Indeed, for a continuous ξ_t we have $p_t(\alpha,\beta] = 0$ if $\alpha\beta > 0$. Hence $\Lambda((\alpha,\beta]) = 0$ and $\varphi(t,\lambda) = \exp\{ibt\lambda - \sigma^2\lambda^2 t/2\}$. For $\sigma \neq 0$, it follows that $\eta_t := (\xi_t - bt)/\sigma$ is a continuous process with independent increments, $\eta_0 = 0$, and $\eta_t - \eta_s \sim N(0, |t-s|)$. As we know, η_t is a Wiener process. If $\sigma = 0$, then $\xi_t - bt = 0$ (a.s.) for any t and, actually, $\xi_t - bt = 0$ for all t at once (a.s.) since $\xi_t - bt$ is continuous.

12. Corollary. *Let an open set $G \subset \mathbb{R}\setminus\{0\}$ be such that $\Lambda(G) = 0$. Then there exists $\Omega' \in \mathcal{F}$ such that $P(\Omega') = 1$ and, for each $t \geq 0$ and $\omega \in \Omega'$, $\Delta\xi_t(\omega) \notin G$.*

Indeed, represent G as a countable union (perhaps with intersections) of intervals $(a_m, b_m]$. Since $\Lambda((a_m, b_m]) = 0$, we have $Ep_t(a_m, b_m] = 0$ and $p_t(a_m, b_m] = 0$ (a.s.). Adding to this that $p_t(a_m, b_m]$ increases in t, we conclude that $p_t(a_m, b_m] = 0$ for all t (a.s.). Now let

$$\Omega' = \bigcap_m \{\omega : p_t(a_m, b_m] = 0 \quad \forall t \geq 0\}.$$

Then $P(\Omega') = 1$ and

$$p((0,t] \times G) \leq \sum_m p_t(a_m, b_m] = 0$$

for each $\omega \in \Omega'$ and $t \geq 0$, as asserted.

The following corollary will be used for deriving an integral representation of ξ_t through jump measures.

13. Corollary. *Denote $q_t(a,b] = p_t(a,b] - t\Lambda((a,b])$. Let some numbers satisfying $a_i \leq b_i$ and $a_i b_i > 0$ be given for $i = 1, 2$. Then, for all $t, s \geq 0$,*

$$Eq_t(a_1, b_1]q_s(a_2, b_2] = (s \wedge t)\Lambda((a_1, b_1] \cap (a_2, b_2]). \tag{12}$$

Indeed, without loss of generality assume $t \geq s$. Notice that both parts of (12) are additive in the sense that if, say, $(a_1, b_1] = (a_3, b_3] \cup (a_4, b_4]$ and $(a_3, b_3] \cap (a_4, b_4] = \emptyset$, then

$$q_t(a_1, b_1] = q_t(a_3, b_3] + q_t(a_4, b_4],$$

$$\Lambda((a_1, b_1] \cap (a_2, b_2]) = \Lambda((a_3, b_3] \cap (a_2, b_2]) + \Lambda((a_4, b_4] \cap (a_2, b_2]).$$

It follows easily that to prove (12) it suffices to prove it only for two cases: (i) $(a_1, b_1] \cap (a_2, b_2] = \emptyset$ and (ii) $a_1 = a_2$, $b_1 = b_2$.

In the first case (12) follows from the independence of the processes $p_\cdot(a_1, b_1]$ and $p_\cdot(a_2, b_2]$ and from (7). In the second case, it suffices to remember that the variance of a random variable having the Poisson distribution with parameter Λ is Λ and use the fact that

$$q_t(a,b] = q_s(a,b] + (q_t(a,b] - q_s(a,b]),$$

where the summands are independent.

We will also use the following theorem, which is closely related to Theorem 9.

14. Theorem. *Take $a > 0$ and define*

$$\eta_t = \int_{[a,\infty)} x\, p_t(dx) + \int_{(-\infty,-a]} x\, p_t(dx). \tag{13}$$

Then:

(i) *the process η_t is infinitely divisible, cadlag, with $\sigma = b = 0$ and Lévy measure $\Lambda(\Gamma \setminus (-a,a))$;*

(ii) *the process $\xi_t - \eta_t$ is infinitely divisible, cadlag, and does not have jumps larger in magnitude than a;*

(iii) *the processes η_t and $\xi_t - \eta_t$ are independent.*

Proof. Assertion (i) is proved like the similar assertion in Theorem 9 on the basis of Lemma 8. Indeed, take a sequence of continuous functions $f_k(x) \to x(1 - I_{(-a,a)}(x))$ such that $f_k(x) = 0$ for $|x| \le a/2$. Then, for any ω,

$$\int_{\mathbb{R}} f_k(x)\, p_t(dx) \to \eta_t. \tag{14}$$

This and Lemma 8 imply that η_t is a homogeneous process with independent increments. That it is cadlag follows from Remark 2. The stochastic continuity of η_t follows from its right continuity and homogeneity as in Lemma 8. To find the Lévy measure of η_t, observe that $|\exp\{i\lambda f_k(x)\} - 1| \le 2I_{|x|\ge a/2}$. By using (9), Lemma 8, and the dominated convergence theorem, we conclude that

$$Ee^{i\lambda\eta_t} = \exp t \int_{\mathbb{R}} (e^{i\lambda x(1-I_{(-a,a)}(x))} - 1)\, \Lambda(dx) = \exp t \int_{\mathbb{R}\setminus(-a,a)} (e^{i\lambda x} - 1)\, \Lambda(dx).$$

In assertion (ii) the fact that $\xi_t - \eta_t$ is an infinitely divisible cadlag process is proved as above on the basis of Lemma 8. The assertion about its jumps is obvious because of Remark 2. Another explanation of the same fact can be obtained from Lemma 8, which implies that

$$Ee^{i(\lambda\eta_t+\alpha\xi_t)}$$

$$= \exp t\{ \int_{\mathbb{R}} (e^{i\lambda x(1-I_{(-a,a)}(x))+i\alpha x} - 1 - i\alpha \sin x)\,\Lambda(dx) + i\alpha b - \alpha^2\sigma^2/2\}$$

$$= \exp t\{ \int_{(-a,a)} (e^{i\alpha x} - 1 - i\alpha \sin x)\,\Lambda(dx)$$

$$+ \int_{\mathbb{R}\setminus(-a,a)} (e^{i(\lambda+\alpha)x} - 1 - i\alpha \sin x)\,\Lambda(dx) + i\alpha b - \alpha^2\sigma^2/2\}, \quad (15)$$

where, for $\lambda = -\alpha$, the expression in the last braces is

$$\int_{(-a,a)} (e^{i\alpha x} - 1 - i\alpha \sin x)\,\Lambda(dx) + i\alpha(b - \int_{\mathbb{R}\setminus(-a,a)} \sin x\,\Lambda(dx)) - \alpha^2\sigma^2/2,$$

which shows that the Lévy measure of $\xi_t - \eta_t$ is concentrated on $(-a,a)$.

To prove (iii), first take $\lambda = \beta - \alpha$ in (15). Then we see that

$$Ee^{i\beta\eta_t+i\alpha(\xi_t-\eta_t)} = e^{tg},$$

where

$$g = \int_{\mathbb{R}\setminus(-a,a)} (e^{i\beta x} - 1)\,\Lambda(dx)$$

$$+ \int_{(-a,a)} (e^{i\alpha x} - 1 - i\alpha \sin x)\,\Lambda(dx) + i\alpha(b - \int_{\mathbb{R}\setminus(-a,a)} \sin x\,\Lambda(dx)) - \alpha^2\sigma^2/2,$$

so that $Ee^{i\beta\eta_t+i\alpha(\xi_t-\eta_t)} = Ee^{i\beta\eta_t}Ee^{i\alpha(\xi_t-\eta_t)}$. Hence, for any t, η_t and $\xi_t - \eta_t$ are independent.

Furthermore, for any constants $\lambda, \alpha \in \mathbb{R}$, the process $\lambda\eta_t + \alpha(\xi_t - \eta_t) = (\lambda - \alpha)\eta_t + \alpha\xi_t$ is a homogeneous process, which is proved as above by using Lemma 8. It follows that the two-dimensional process $(\eta_t, \xi_t - \eta_t)$ has homogeneous increments. In particular, if $s < t$, the distributions of $(\eta_{t-s}, \xi_{t-s} - \eta_{t-s})$ and $(\eta_t - \eta_s, \xi_t - \eta_t - (\xi_s - \eta_s))$ coincide, and since the first pair is independent, so is the second. Now the independence of the processes η_t and $\xi_t - \eta_t$ follows from Lemma 7 and from the fact that $\eta_t - \eta_s$, $\xi_t - \xi_s$, and $(\eta_t - \eta_s, \xi_t - \eta_t - (\xi_s - \eta_s))$ are $\mathcal{F}^{\xi}_{s,t}$-measurable (see (14) and Lemma 8). The theorem is proved.

The following exercise describes all nonnegative infinitely divisible cadlag processes.

15. Exercise. Let ξ_t be an infinitely divisible cadlag process satisfying $\xi_t \geq 0$ for all $t \geq 0$ and ω. Take $\eta_t = \eta_t(a)$ from Theorem 14.

(i) By using Exercise 2.9, show that all jumps of ξ_t are nonnegative.

(ii) Prove that for every $t \geq 0$, we have $P(\eta_t(a) = 0) = \exp(-t\Lambda([a,\infty)))$.

(iii) From Theorem 14 and (ii), derive that $\xi_t - \eta_t(a) \geq 0$ (a.s.) for each $t \geq 0$.

(iv) Since obviously $\eta_t(a)$ increases as a decreases, conclude that $\eta_t(0+) \leq \xi_t < \infty$ (a.s.) for each $t \geq 0$. From (15) with $\alpha = 0$ find the characteristic function of $\eta_t(0+)$ and prove that $\xi_t - \eta_t(0+)$ has normal distribution. By using that $\xi_t - \eta_t(0+) \geq 0$ (a.s.), prove that $\xi_t = \eta_t(0+)$ (a.s.).

(v) Prove that

$$\int_0^1 x\Lambda(dx) < \infty, \quad \xi_t = \int_{(0,\infty)} x\, p(t,dx) \quad \text{(a.s.)},$$

and, in particular, ξ_t is a pure jump process with nonnegative jumps.

4. Further comments on jump measures

1. Exercise. Let $f(t,x)$ be a Borel nonnegative function such that $f(t,0) = 0$. Prove that $\int_{\mathbb{R}_+\times\mathbb{R}} f(s,x)\,p(dsdx)$ is a random variable and

$$E\int_{\mathbb{R}_+\times\mathbb{R}} f(s,x)\,p(dsdx) = \int_{\mathbb{R}_+\times\mathbb{R}} f(s,x)\,ds\Lambda(dx). \tag{1}$$

2. Exercise. Let $f(t,x) = f(\omega,t,x)$ be a bounded function such that $f = 0$ for $|x| < \varepsilon$ and for $t \geq T$, where the constants $\varepsilon, T \in (0,\infty)$. Also assume that $f(\omega,t,x)$ is left continuous in t for any (ω,x) and $\mathcal{F}_t^\xi \otimes \mathfrak{B}(\mathbb{R})$-measurable for any t. Prove that the following version of (1) holds:

$$E\int_{\mathbb{R}_+\times\mathbb{R}} f(s,x)\,p(dsdx) = \int_{\mathbb{R}_+\times\mathbb{R}} Ef(s,x)\,ds\Lambda(dx).$$

The following two exercises are aimed at generalizing Theorem 3.9.

3. Exercise. Let $f(t,x)$ be a bounded Borel function such that $f = 0$ for $|x| < \varepsilon$, where the constant $\varepsilon > 0$. Prove that, for $t \in [0,\infty)$,

$$\varphi(t) := E\exp\{i\int_{(0,t]\times\mathbb{R}} f(s,x)\,p(dsdx)\} = \exp\{\int_{(0,t]\times\mathbb{R}} (e^{if(s,x)}-1)\,ds\Lambda(dx)\}.$$

4. Exercise. By taking f in Exercise 3 as linear combinations of the indicators of Borel subsets $\Gamma_1, ..., \Gamma_n$ of $\mathbb{R}_+ \times \mathbb{R}$, prove that, if the sets are disjoint, then $p(\Gamma_1), ..., p(\Gamma_n)$ are independent. Also prove that, if $\Gamma_1 \subset R_{T,\varepsilon}$, then $p(\Gamma_1)$ is Poisson with parameter $(\ell \times \Lambda)(\Gamma_1)$.

The following exercise shows that Poisson processes without common jumps are independent.

5. Exercise. Let $(\Omega, \mathcal{F}, P)$ be a probability space, and let $\mathcal{F}_t$ be σ-fields defined for $t \geq 0$ such that $\mathcal{F}_s \subset \mathcal{F}_t \subset \mathcal{F}$ for $s \leq t$. Assume that ξ_t and η_t are two Poisson processes with parameters μ and ν respectively defined on Ω, and such that ξ_t and η_t are $\mathcal{F}$-measurable for each t and $\xi_{t+h} - \xi_t$ and $\eta_{t+h} - \eta_t$ are independent of $\mathcal{F}_t$ for all $t, h \geq 0$. Finally, assume that ξ_t and η_t do not have common jumps, that is, $(\Delta\xi_t)\Delta\eta_t = 0$ for all t and ω. Prove that the processes ξ_t and η_t are independent.

5. Representing infinitely divisible processes through jump measures

We start with a simple result.

1. Theorem. *Let ξ_t be an infinitely divisible cadlag process with parameters σ, b, and Lévy measure concentrated at points $x_1, ..., x_n$.*

(i) *If $\sigma \neq 0$, then there exist a Wiener process w_t and Poisson processes $p_t^1, ..., p_t^n$ with parameters $\Lambda(\{x_1\}), ..., \Lambda(\{x_n\})$, respectively, such that $w_t, p_t^1, ..., p_t^n$ are mutually independent and*

$$\xi_t = x_1 p_t^1 + ... + x_n p_t^n + bt + \sigma w_t \quad \forall t \geq 0 \quad (a.s.). \tag{1}$$

(ii) *If $\sigma = 0$, assertion* (i) *still holds if one does not mention w_t and drops the term σw_t in* (1).

Proof. (i) Of course, we assume that $x_i \neq x_j$ for $i \neq j$. Notice that $\Lambda(\{0\}) = 0$. Therefore, $x_m \neq 0$. Also

$$\Lambda(\mathbb{R} \setminus \{x_1, ..., x_n\}) = 0.$$

Hence, by Corollary 3.12, we may assume that all jumps of ξ_t belong to the set $\{x_1, ..., x_n\}$.

Now take $a > 0$ such that $a < |x_i|$ for all i, and define η_t by (3.13). By Theorem 3.14 the process $\xi_t - \eta_t$ does not have jumps and is infinitely divisible. By Corollary 3.11 we conclude that

$$\xi_t - \eta_t = bt + \sigma w_t.$$

In addition, formula (3.1) shows also that

$$\eta_t = x_1 p_t(\{x_1\}) + ... + x_n p_t(\{x_n\}) = x_1 p_t(a_1, b_1] + ... + x_n p_t(a_n, b_n],$$

where a_m, b_m are any numbers satisfying $a_m b_m > 0$, $a_m < x_m \le b_m$, and such that $(a_m, b_m]$ are mutually disjoint. This proves (1) with $p_t^m = p_t(a_m, b_m]$, which are Poisson processes with parameters $\Lambda(\{x_m\})$.

To prove that $w_t, p_t^1, ..., p_t^n$ are mutually independent, introduce p^η as the jump measure of η_t and observe that by Theorem 3.14 the processes $\xi_t - \eta_t = bt + \sigma w_t$ and η_t (that is, w_t and η_t) are independent. It follows from Lemma 3.3 that, if we take any continuous functions $f_1, ..., f_n$ vanishing in the neighborhood of the origin, then the process w_t and the vector-valued process

$$\Big(\int_{\mathbb{R}} f_1(x)\, p_t^\eta(dx), ..., \int_{\mathbb{R}} f_n(x)\, p_t^\eta(dx)\Big)$$

are independent. By taking appropriate approximations we conclude that the process w_t and the vector-valued process

$$(p_t^\eta(a_1, b_1], ..., p_t^\eta(a_n, b_n])$$

are independent. Finally, by observing that, by Theorem 3.9, the processes $p_t^\eta(a_1, b_1]$,..., $p_t^\eta(a_n, b_n]$ are independent and, obviously (cf. (3.2)), $p^\eta = p$, we get that $w_t, p_t^1, ..., p_t^n$ are mutually independent. The theorem is proved.

The above proof is based on the formula

$$\xi_t = \zeta_t^a + \eta_t^a, \tag{2}$$

where

$$\eta_t^a = \int_{\mathbb{R}\setminus(-a,a)} x\, p_t(dx), \quad \zeta_t^a = \xi_t - \eta_t^a, \quad a > 0,$$

and the fact that for small a all processes η_t^a are the same. In the general case we want to let $a \downarrow 0$ in (2). The only trouble is that generally there is no limit of η_t^a as $a \downarrow 0$. On the other hand, the left-hand side of (2) does have a limit, just because it is independent of a. So there is a hope that if we subtract an appropriate quantity from ζ_t^a and add it to η_t^a, the results will converge. This appropriate quantity turns out to be the stochastic integral against *the centered Poisson measure* q introduced by

$$q_t(a, b] = p_t(a, b] - t\Lambda((a, b]) \quad \text{if} \quad ab > 0.$$

2. Lemma. *Let* $\Pi = \{(0,t] \times (a,b] : t > 0, a < b, ab > 0\}$ *and for* $A = (0,t] \times (a,b] \in \Pi$ *let* $q(A) = q_t(a,b]$. *Then* Π *is a* π*-system and* q *is a random orthogonal measure on* Π *with reference measure* $\ell \times \Lambda$.

Proof. Let $A = (0,t_1] \times (a_1,b_1], B = (0,t_2] \times (a_2,b_2] \in \Pi$. Then

$$AB = (0, t_1 \wedge t_2] \times (c,d], \quad (c,d] := (a_1,b_1] \cap (a_2,b_2],$$

which shows that Π is a π-system. That q is a random orthogonal measure on Π with reference measure $\ell \times \Lambda$ is stated in Corollary 3.13. The lemma is proved.

3. Remark. We may consider Π as a system of subsets of $\mathbb{R}_+ \times \mathbb{R} \setminus \{0\}$. Then as is easy to see, $\sigma(\Pi) = \mathfrak{B}(\mathbb{R}_+) \otimes \mathfrak{B}(\mathbb{R} \setminus \{0\})$. By Theorem 2.3.19, $L_2(\Pi, \Lambda) = L_2(\sigma(\Pi), \ell \times \Lambda)$. Therefore, Lemma 2 and Theorem 2.3.13 allow us to define the stochastic integral

$$\int_{\mathbb{R}_+ \times (\mathbb{R} \setminus \{0\})} f(t,x)\, q(dtdx)$$

for every Borel f satisfying

$$\int_{\mathbb{R}_+ \times \mathbb{R}} |f(t,x)|^2\, dt\Lambda(dx) < \infty$$

(we write this integral over $\mathbb{R}_+ \times \mathbb{R}$ instead of $\mathbb{R}_+ \times (\mathbb{R} \setminus \{0\})$ because $\Lambda(\{0\}) = 0$ by definition). Furthermore,

$$\begin{aligned} E\Big|\int_{\mathbb{R}_+ \times (\mathbb{R} \setminus \{0\})} f(t,x)\, q(dtdx)\Big|^2 &= \int_{\mathbb{R}_+ \times \mathbb{R}} |f(t,x)|^2\, dt\Lambda(dx), \\ E\int_{\mathbb{R}_+ \times (\mathbb{R} \setminus \{0\})} f(t,x)\, q(dtdx) &= 0, \end{aligned} \tag{3}$$

the latter following from the fact that $Eq(A) = 0$ if $A \in \Pi$ (see Remark 2.3.15).

4. Remark. Denote

$$\int_{\mathbb{R} \setminus \{0\}} f(x)\, q_t(dx) = \int_{\mathbb{R}_+ \times (\mathbb{R} \setminus \{0\})} I_{(0,t]}(u) f(x)\, q(dudx). \tag{4}$$

Then (3) shows that, for each Borel f satisfying $\int_{\mathbb{R}} |f(x)|^2\, \Lambda(dx) < \infty$ and every $t, s \in [0, \infty)$,

$$E\Big|\int_{\mathbb{R}\setminus\{0\}} f(x)\,q_t(dx)-\int_{\mathbb{R}\setminus\{0\}} f(x)\,q_s(dx)\Big|^2=|t-s|\int_{\mathbb{R}}|f(x)|^2\,\Lambda(dx),$$

$$E\Big|\int_{\mathbb{R}\setminus\{0\}} f(x)\,q_t(dx)\Big|^2=t\int_{\mathbb{R}}|f(x)|^2\,\Lambda(dx).$$

In the following exercise we use for the first time our assumption that $(\Omega,\mathcal{F},P)$ is a complete probability space. This assumption allowed us to complete $\sigma(\xi_s:s\le t)$ and have this completion, denoted $\mathcal{F}^\xi_t$, to be part of $\mathcal{F}$. This assumption implies that, if we are given two random variables satisfying $\zeta=\eta$ (a.s) and ζ is $\mathcal{F}^\xi_t$-measurable, so is η.

5. Exercise*. Prove that if f is a bounded Borel function vanishing in a neighborhood of zero, then $\int_{\mathbb{R}}|f(x)|^2\,\Lambda(dx)<\infty$ and

$$\int_{\mathbb{R}\setminus\{0\}} f(x)\,q_t(dx)=\int_{\mathbb{R}} f(x)\,p_t(dx)-t\int_{\mathbb{R}} f(x)\,\Lambda(dx)\quad\text{(a.s.).}\tag{5}$$

By using Lemma 3.8, conclude that the left-hand side of (5) is $\mathcal{F}^\xi_t$-measurable for every $f\in L_2(\mathfrak{B}(\mathbb{R}),\Lambda)$.

6. Exercise*. As a continuation of Exercise 5, prove that (5) holds for every Borel f satisfying $f(0)=0$ and $\int_{\mathbb{R}}(|f|+|f|^2)\,\Lambda(dx)<\infty$.

7. Lemma. *For every Borel $f\in L_2(\mathfrak{B}(\mathbb{R}),\Lambda)$ the stochastic integral*

$$\eta_t:=\int_{\mathbb{R}\setminus\{0\}} f(x)\,q_t(dx)$$

is an infinitely divisible $\mathcal{F}^\xi_t$-adapted process such that, if $0\le s\le t<\infty$, then $\eta_t-\eta_s$ and $\mathcal{F}^\xi_s$ are independent. By Theorem 1.11 the process η_t admits a modification with trajectories in $D[0,\infty)$. If we keep the same notation for the modification, then for every $T\in[0,\infty)$

$$E\sup_{t\le T}\eta_t^2\le 4T\int_{\mathbb{R}}|f(x)|^2\,\Lambda(dx).\tag{6}$$

Proof. If f is a bounded continuous function vanishing in a neighborhood of zero, the first statement follows from Exercise 5 and Lemma 3.8. An obvious approximation argument and Remark 4 allow us to extend the result to arbitrary f in question.

To prove (6) take $0\le t_1\le\ldots\le t_n\le T$ and observe that, owing to the independence of $\eta_{t_{k+1}}-\eta_{t_k}$ and $\mathcal{F}^\xi_{t_k}$, we have

$$E(\eta_{t_{k+1}} - \eta_{t_k} | \mathcal{F}^{\xi}_{t_k}) = E(\eta_{t_{k+1}} - \eta_{t_k}) = 0.$$

Therefore, $(\eta_{t_k}, \mathcal{F}^{\xi}_{t_k})$ is a martingale. By Doob's inequality

$$E \sup_k \eta^2_{t_k} \le 4E\eta^2_T = 4T \int_{\mathbb{R}} |f(x)|^2 \, \Lambda(dx).$$

Clearly the inequality between the extreme terms has nothing to do with ordering t_k. Therefore by ordering the set ρ of all rationals on $[0,T]$ and taking the first n rationals as t_k, $k = 1, ..., n$, and then sending n to infinity, by Fatou's theorem we find that

$$E \sup_{r \in \rho, r < T} \eta^2_r \le 4T \int_{\mathbb{R}} |f(x)|^2 \, \Lambda(dx).$$

Now equation (6) immediately follows from the right continuity and the stochastic continuity (at point T) of $\eta_\cdot$, since (a.s.)

$$\sup_{t \le T} \eta^2_t = \sup_{t < T} \eta^2_t = \sup_{r \in \rho, r < T} \eta^2_r.$$

The lemma is proved.

8. Theorem. *Let ξ_t be an infinitely divisible cadlag process with parameters σ, b, and Lévy measure Λ.*

(i) If $\sigma \ne 0$, then there exist a constant $\bar{b}$ and a Wiener process w_t, which is independent of all processes $p_t(c,d]$, such that, for each $t \ge 0$,

$$\xi_t = \bar{b}t + \sigma w_t + \int_{(-1,1)} x \, q_t(dx) + \int_{\mathbb{R} \setminus (-1,1)} x \, p_t(dx) \quad (a.s.). \tag{7}$$

(ii) If $\sigma = 0$, assertion (i) still holds if one does not mention w_t and drops the term σw_t in (7).

Proof. For $a \in (0,1)$ write (2) as

$$\xi_t = \zeta^a_t + \int_{(-1,1) \setminus (-a,a)} x \, p_t(dx) + \int_{\mathbb{R} \setminus (-1,1)} x \, p_t(dx).$$

Here, by Exercise 5,

$$\int_{(-1,1) \setminus (-a,a)} x \, p_t(dx) = \int_{(-1,1) \setminus (-a,a)} x \, q_t(dx) + t \int_{(-1,1) \setminus (-a,a)} x \, \Lambda(dx),$$

so that

$$\xi_t = \kappa_t^a + \int_{(-1,1)\setminus(-a,a)} x\, q_t(dx) + \int_{\mathbb{R}\setminus(-1,1)} x\, p_t(dx), \tag{8}$$

where

$$\kappa_t^a = \zeta_t^a + t \int_{(-1,1)\setminus(-a,a)} x\, \Lambda(dx).$$

By Lemma 7, for any $T \in (0, \infty)$,

$$E \sup_{t \le T} \Big| \int_{(-1,1)\setminus(-a,a)} x\, q_t(dx) - \int_{(-1,1)} x\, q_t(dx) \Big|^2 \to 0$$

as $a \downarrow 0$. Therefore, there exists a sequence $a_n \downarrow 0$, along which with probability one the first integral on the right in (8) converges uniformly on each finite time interval to the first integral on the right in (7). It follows from (8) that almost surely $\kappa_t^{a_n}$ also converges uniformly on each finite time interval to a process, say κ_t. Bearing in mind that the κ_t^a are cadlag and using Exercise 1.8, we see that κ_t is cadlag too. By Theorem 3.14, the process ζ_t^a is infinitely divisible cadlag. It follows that κ_t^a and κ_t are infinitely divisible cadlag as well.

Furthermore, since ζ_t^a does not have jumps larger in magnitude than a, the process κ_t does not have jumps at all and hence is continuous (the last conclusion is easily proved by contradiction). Again by Theorem 3.14, the process ζ_t^a is independent of η_t^a and, in particular, is independent of the jump measure of η_t^a (cf. Lemma 3.3). The latter being $p_t((c,d]\setminus(-a,a))$ (cf. (3.2)) shows that ζ_t^a as well as κ_t^a are independent of all processes $p_t((c,d]\setminus(-a,a))$. By letting $a \downarrow 0$, we conclude that κ_t is independent of all processes $p_t(c,d]$.

To conclude the proof it only remains to use Corollary 3.11. The theorem is proved.

9. Exercise. It may look as though assertion (i) of Theorem 8 holds even if $\sigma = 0$. Indeed, in this case $\sigma w_t \equiv 0$ anyway. However, generally this assertion is false if $\sigma = 0$. The reader is asked to give an example in which this happens.

6. Constructing infinitely divisible processes

Here we want to show that for an arbitrary Lévy measure and constants b and σ there exists an infinitely divisible process ξ_t, defined on an appropriate probability space, such that

$$Ee^{i\lambda\xi_t} = \exp t\{\int_{\mathbb{R}} (e^{i\lambda x} - 1 - i\lambda \sin x)\,\Lambda(dx) + ib\lambda - \sigma^2\lambda^2/2\}. \tag{1}$$

By the way, this will show that generally there are no *additional* properties of Λ apart from those listed in (2.7).

The idea is that if we have at least one process with "arbitrarily" small jumps, then by "redirecting" the jumps we can get jump measures corresponding to arbitrary infinitely divisible process. We know that at least one such "test" process exists, the increasing 1/2-stable process τ_{a+}, $a \geq 0$ (see Theorem 2.6.1 and Exercises 1.12).

The following lemma shows how to redirect the jumps of τ_{a+}.

1. Lemma. *Let Λ be a positive measure on $\mathfrak{B}(\mathbb{R})$ such that $\Lambda(\mathbb{R}\setminus(-a,a)) < \infty$ for any $a > 0$ and $\Lambda(\{0\}) = 0$. Then there exists a finite Borel function $f(x)$ on $\mathbb{R}$ such that $f(0) = 0$ and for any Borel Γ*

$$\Lambda(\Gamma) = \int_{f^{-1}(\Gamma)} |x|^{-3/2}\,dx.$$

Proof. For $x > 0$, define $2F(x) = \Lambda\{(x,\infty)\}$. Notice that $F(x)$ is right continuous on $(0,\infty)$ and $F(\infty) = 0$. For $x > 0$ let

$$f(x) = \inf\{y > 0 : 1 \geq xF^2(y)\}.$$

Since $F(\infty) = 0$, f is a finite function.

Next notice that, if $t > 0$ and $f(x) > t$, then for any $y > 0$ satisfying $1 \geq xF^2(y)$, we have $y > t$, which implies that $1 < xF^2(t)$. Hence,

$$\{x > 0 : f(x) > t\} \subset \{x > 0 : xF^2(t) > 1\}. \tag{2}$$

On the other hand, if $t > 0$ and $xF^2(t) > 1$, then due to the right continuity of F also $xF^2(t+\varepsilon) > 1$, where $\varepsilon > 0$. In that case, $f(x) \geq t + \varepsilon > t$. Thus the sets in (2) coincide if $t > 0$, and hence

$$\Lambda\{(t,\infty)\} = 2F(t) = \int_{1/F^2(t)}^{\infty} x^{-3/2}\,dx = \int_{x:xF^2(t)>1} x^{-3/2}\,dx = \nu\{(t,\infty)\},$$

where

$$\nu(\Gamma) = \int_{x>0:f(x)\in\Gamma} x^{-3/2}\,dx.$$

A standard measure-theoretic argument allows us to conclude that

$$\Lambda(\Gamma \cap (0, \infty)) = \nu(\Gamma)$$

not only for $\Gamma = (t, \infty)$, $t > 0$, but for all Borel $\Gamma \subset (0, \infty)$.

Similarly, one constructs a negative function $g(x)$ on $(-\infty, 0)$ such that

$$\Lambda(\Gamma \cap (-\infty, 0)) = \int_{x<0: g(x) \in \Gamma} |x|^{-3/2}\, dx.$$

Finally, the function we need is given by $f(x)I_{x>0} + g(x)I_{x<0}$. The lemma is proved.

We also need the following version of Lemma 3.8.

2. Lemma. *Let p_t be the jump measure of an infinitely divisible cadlag process with Lévy measure Λ, and let f be a finite Borel function such that $f(0) = 0$ and $\Lambda(\{x : f(x) \neq 0\}) < \infty$. Then*

(i) *we have*

$$\int_{\mathbb{R}\setminus\{0\}} |f(x)|\, p_t(dx) < \infty$$

(a.s.), and

$$\xi_t := \int_{\mathbb{R}\setminus\{0\}} f(x)\, p_t(dx)$$

is well defined and is cadlag;

(ii) *ξ_t is an infinitely divisible process, and*

$$Ee^{i\xi_t} = \exp t \int_{\mathbb{R}} (e^{if(x)} - 1)\, \Lambda(dx). \tag{3}$$

Proof. (i) By Corollary 3.10

$$Ep_t(\{x : f(x) \neq 0\}) = t\Lambda(\{x : f(x) \neq 0\}) < \infty.$$

Since the measure p_t is integer valued, it follows that (a.s.) there are only finitely many points in $\{x : f(x) \neq 0\}$ to which p_t assigns a nonzero mass. This proves (i).

To prove (ii) we use approximations. The inequality $|e^{if} - 1| \leq 2I_{f\neq 0}$ and the dominated convergence theorem show that, if assertion (ii) holds for some functions $f_n(x)$ such that $f_n \xrightarrow{\Lambda} f$, $\Lambda(\{x : \sup_n |f_n(x)| > 0\}) < \infty$, and

$$\int_{\mathbb{R}\setminus\{0\}} |f - f_n|\, p_t(dx) \xrightarrow{P} 0, \tag{4}$$

then (ii) is also true for f. By taking $f_n = (-n) \vee f \wedge n$, we see that it suffices to prove (ii) for bounded f. Then considering $f_n = fI_{1/n<|x|<n}$ reduces the general case further to bounded functions vanishing for small and large $|x|$. Any such function can be approximated in $L_1(\mathfrak{B}(\mathbb{R}), \Lambda)$ by continuous functions f_n, for which (4) holds automatically due to Corollary 3.10 and (3) holds due to Lemma 3.8 (ii) with $\alpha = 0$. The lemma is proved.

Now let Λ be a Lévy measure and b and σ some constants. Take a probability space carrying two independent copies $\eta_t^{\pm}$ of the process τ_{t+}, $t \geq 0$, and a Wiener process w_t independent of (η_t^+, η_t^-). By Exercise 2.11, the Lévy measure of $\eta_t^{\pm}$ is given by $c_0 x^{-3/2} I_{x>0}\, dx$, where c_0 is a constant. Define

$$\Lambda_0(dx) = c_0|x|^{-3/2}\, dx$$

and take the function f from Lemma 1 constructed from Λ/c_0 in place of Λ, so that, for any $\Gamma \in \mathfrak{B}(\mathbb{R})$,

$$\Lambda(\Gamma) = \Lambda_0 f^{-1}(\Gamma) = \Lambda_0(\{x : f(x) \in \Gamma\}). \tag{5}$$

3. Remark. Equation (5) means that, for any $\Gamma \in \mathfrak{B}(\mathbb{R})$ and $h = I_\Gamma$,

$$\int_{\mathbb{R}} h(x)\, \Lambda(dx) = \int_{\mathbb{R}} h(f(x))\, \Lambda_0(dx). \tag{6}$$

A standard measure-theoretic argument shows that (6) is true for each Borel nonnegative h and also for each Borel h for which at least one of

$$\int_{\mathbb{R}} |h(x)|\, \Lambda(dx) \quad \text{and} \quad \int_{\mathbb{R}} |h(f(x))|\, \Lambda_0(dx)$$

is finite. In particular, if h is a Borel function, then $h \in L_2(\mathfrak{B}(\mathbb{R}), \Lambda)$ if and only if $h(f) \in L_2(\mathfrak{B}(\mathbb{R}), \Lambda_0)$.

4. Theorem. *Let $p^{\pm}$ be the jump measures of $\eta_t^{\pm}$ and $q^{\pm}$ the centered Poisson measures of $\eta_t^{\pm}$. Define*

$$\xi_t^\pm = \int_{\mathbb{R}\setminus\{0\}} f(\pm x) I_{|f(\pm x)|<1}\, q_t^\pm(dx) + \int_{\mathbb{R}\setminus\{0\}} f(\pm x) I_{|f(\pm x)|\geq 1}\, p_t^\pm(dx)$$
$$=: \alpha_t^\pm + \beta_t^\pm .$$

Then, for a constant $\bar{b}$, the process

$$\xi_t = \bar{b}t + \sigma w_t + \xi_t^+ + \xi_t^-$$

is an infinitely divisible process satisfying (1).

Proof. Observe that

$$\int_{\mathbb{R}} f^2(\pm x) I_{|f(\pm x)|<1}\, \Lambda_0(dx) = \int_{(-1,1)} x^2\, \Lambda(dx) < \infty.$$

Therefore, the processes $\alpha_t^\pm$ are well defined. In addition,

$$\Lambda_0(\{x > 0 : |f(\pm x)| \geq 1\}) \leq \Lambda_0(\{x : |f(x)| \geq 1\}) = \Lambda(|x| \geq 1) < \infty.$$

Hence, $\beta_t^\pm$ is well defined due to Lemma 2.

Next, in order to find the characteristic function of $\xi_t^\pm$, notice that $f I_{|f|<a} \to 0$ in $L_2(\mathfrak{B}(\mathbb{R}), \Lambda_0)$ as $a \downarrow 0$, so that upon remembering the properties of stochastic integrals, in particular, Exercise 5.6, we obtain

$$\alpha_t^\pm = \underset{a\downarrow 0}{\text{l.i.m.}} \Big(\int_{\mathbb{R}\setminus\{0\}} f(\pm x) I_{a\leq|f(\pm x)|<1}\, p_t^\pm(dx) - t\int_0^\infty f(\pm x) I_{a\leq|f(\pm x)|<1}\, \Lambda_0(dx)\Big).$$

It follows that

$$\xi_t^\pm = P\text{-}\lim_{a\downarrow 0} \Big(\int_{\mathbb{R}\setminus\{0\}} f(\pm x) I_{a\leq|f(\pm x)|}\, p_t^\pm(dx) - t\int_0^\infty f(\pm x) I_{a\leq|f(\pm x)|<1}\, \Lambda_0(dx)\Big).$$

Now Lemma 2 implies that $\xi_t^\pm$ are infinitely divisible and

$$Ee^{i\lambda\xi_t^\pm} = \lim_{a\downarrow 0} \exp t\int_0^\infty \big\{(e^{i\lambda f(\pm x)} - 1) I_{a\leq|f(\pm x)|} - i\lambda f(\pm x) I_{a\leq|f(\pm x)|<1}\big)\big\}\, \Lambda_0(dx).$$

In the next few lines we use the fact that $|e^{i\lambda x} - 1 - i\lambda x I_{|x|<1}|$ is less than $\lambda^2 x^2$ if $|x| < 1$ and less than 2 otherwise. Then, owing to Remark 3, we find that

$$g(\lambda, a) := \int_0^\infty \{(e^{i\lambda f(x)} - 1)I_{a\le|f(x)|} - i\lambda f(x) I_{a\le|f(x)|<1}\}\,\Lambda_0(dx)$$

$$+\int_0^\infty \{(e^{i\lambda f(-x)} - 1)I_{a\le|f(-x)|} - i\lambda f(-x) I_{a\le|f(-x)|<1}\}\,\Lambda_0(dx)$$

$$= \int_{\mathbb{R}} \{(e^{i\lambda f(x)} - 1)I_{a\le|f(x)|} - i\lambda f(x) I_{a\le|f(x)|<1}\}\,\Lambda_0(dx)$$

$$= \int_{\mathbb{R}\setminus(-a,a)} \{e^{i\lambda x} - 1 - i\lambda x I_{|x|<1}\}\,\Lambda(dx).$$

This along with the dominated convergence theorem implies that

$$g(\lambda, a) \to \int_{\mathbb{R}} (e^{i\lambda x} - 1 - i\lambda x I_{|x|<1})\,\Lambda(dx) = \int_{\mathbb{R}} (e^{i\lambda x} - 1 - i\lambda \sin x)\,\Lambda(dx) + i\lambda\tilde{b},$$

where

$$\tilde{b} = \int_{\mathbb{R}} (\sin x - x I_{|x|<1})\,\Lambda(dx)$$

is a well-defined constant because $|\sin x - x I_{|x|<1}| \le 2 \wedge x^2$.

Finally, upon remembering that the processes w_t, p_t^+, p_t^- are independent, we conclude that ξ_t is infinitely divisible and

$$Ee^{i\lambda\xi_t} = \lim_{a\downarrow 0} \exp t\big(i\lambda\bar{b} - \sigma^2\lambda^2/2 + g(\lambda, a)\big),$$

which equals the right-hand side of (1) if $\bar{b} + \tilde{b} = b$. The theorem is proved.

The theory in this chapter admits a very natural generalization for vector-valued infinitely divisible processes, which are defined in the same way as in Sec. 2. Also as above, having an infinitely divisible process with jumps of all sizes in all directions allows one to construct all other infinitely divisible processes. In connection with this we set the reader the following.

5. Exercise. Let $w_t, w_t^1, ..., w_t^d$ be independent Wiener processes. Define $\tau_t = \inf\{s \ge 0 : w_s \ge t\}$ and

$$\eta_t = (w_t^1, ..., w_t^d), \quad \xi_t = \eta_{\tau_t}.$$

Prove that:

(i) The process ξ_t is infinitely divisible.

(ii) $E\exp(i\lambda\cdot\xi_t) = \exp(-ct|\lambda|)$ for any $\lambda \in \mathbb{R}^d$, where $c > 0$ is a constant, so that ξ_t has a multidimensional Cauchy distribution.

(iii) It follows from (ii) that the components of ξ_t are not independent. On the other hand, the components of η_t are independent random processes and we do a change of time in η_t, random but yet independent of η. Explain why this makes the components of $\xi_t = \eta_{\tau_t}$ depend on each other. What kind of nontrivial information about the trajectory of ξ_t^2 can one get if one knows the trajectory $\xi_t^1, t > 0$?

7. Hints to exercises

1.8 Assume the contrary.

1.12 For any cadlag modification $\tilde{\xi}_t$ of a process ξ_t we have $\xi_t \xrightarrow{P} \tilde{\xi}_s$ as $t \downarrow s$.

2.10 Use $\int_{\mathbb{R}}(\lambda\sin(x/\lambda) - \sin x)x^{-2}\,dx = 0$, which is true since $\sin x$ is an odd function.

2.11 To show that $a = b = 1$, observe that

$$\Psi(z) := \int_0^\infty x^{-3/2}(e^{-zx} - 1)\,dx$$

is an analytic function for $\operatorname{Re} z > 0$ which is continuous for $\operatorname{Re} z \geq 0$. Furthermore, for real z, changing variables, prove that $\Psi(z) = \sqrt{z}\Psi(1)$ and express $\Psi(1)$ through the gamma function by integrating by parts. Then notice that $\sqrt{2\pi}\Psi(i) = -a - ib$.

3.15 (ii) $P(\eta_t(a) = 0) = P(p[a,\infty) = 0)$. (iii) Use that $\xi_t - \eta_t(a)$ and $\eta_t(a)$ are independent and their sum is positive. (iv)&(v) Put $\alpha = 0$ in (3.15) to get the characteristic function of $\eta_t(0+)$ and also the fact that

$$\lim_{a\downarrow 0}\int_{[a,\infty)}(e^{i\lambda x} - 1)\,\Lambda(dx)$$

exists.

4.1 Corollary 3.10 says that the finite measures

$$\nu_{\varepsilon,T}(\Gamma) := Ep\{((0,T] \times (\mathbb{R} \setminus (-\varepsilon,\varepsilon))) \cap \Gamma\}$$

and

$$(\ell \times \Lambda)\{((0,T] \times (\mathbb{R} \setminus (-\varepsilon,\varepsilon))) \cap \Gamma\}$$

coincide on sets Γ of the form $(0,t] \times (a,b]$.

4.2 Assume $f \geq 0$, approximate f by the functions $f([tn]/n, x)$, and prove that

$$E\int_{(k/n,(k+1)/n]\times\mathbb{R}} f(k/n,x)\,p(dsdx)$$

$$= E\int_{(k/n,(k+1)/n]\times\mathbb{R}} (Ef(k/n,x))\,p(dsdx).$$

To do this step, use π- and λ-systems in order to show that it suffices to take $f(k/n,x)$ equal to $I_{A\times\Gamma}(\omega,x)$, where A and $p_{(k+1)/n}(\Gamma)-p_{k/n}(\Gamma)$ are independent.

4.3 First let there be an integer n such that $f(t,x)=f((k+1)2^{-n},x)$ whenever k is an integer and $t\in(k2^{-n},(k+1)2^{-n}]$, and let $f((k+1)2^{-n},x)$ be continuous in x. In that case use Lemma 3.8. Then derive the result for any continuous function $f(t,x)$ vanishing for $|x|<\varepsilon$. Finally, pass to the limit from continuous functions to arbitrary ones by using (4.1).

4.5 Take some constants α and β and define $\zeta_t=\alpha\xi_t+\beta\eta_t$, $\varphi(t)=Ee^{i\zeta_t}$. Notice that

$$e^{i\zeta_t}=1+\int_{(0,t]} e^{i\zeta_{t-}}\{[e^{i\alpha}-1]\,d\xi_t+[e^{i\beta}-1]\,d\eta_t\},$$

where on the right we just have a telescoping sum. By taking expectations derive that

$$\varphi(t)=1+\int_0^t \varphi(s)\{[e^{i\alpha}-1]\mu+[e^{i\beta}-1]\nu\}\,ds.$$

This will prove the independence of ξ_t and η_t for any t. To prove the independence of the processes, repeat part of the proof of Lemma 3.7.

5.5 First check (5.5) for $f=I_{(a,b]}$ with $ab>0$, and then use Corollary 3.10, the equality $L_2(\Pi,\Lambda)=L_2(\sigma(\Pi),\Lambda)$, and (2.7), which shows that $\Lambda(\mathbb{R}\setminus(-a,a))<\infty$ for any $a>0$.

5.6 The functions $(n\wedge f)I_{|x|>1/n}$ converge to f in $L_1(\mathfrak{B}(\mathbb{R}),\Lambda)$ and in $L_2(\mathfrak{B}(\mathbb{R}),\Lambda)$.

6.5 (i) Use Theorem 2.6.1. (ii) Add that

$$E\exp(i\lambda\cdot\xi_t)=\int_0^\infty E\exp(i\lambda\cdot\eta_s)\,P(\tau_t\in ds).$$

(iii) Think of jumps.

Chapter 6

Itô Stochastic Integral

The reader may have noticed that stochastic integrals or stochastic integral equations appear in every chapter in this book. Here we present a systematic study of the Itô stochastic integral against the Wiener process. This integral has already been introduced in Sec. 2.7 by using an approach which is equally good for defining stochastic integrals against martingales. This approach also exhibits the importance of the σ-field of predictable sets. Traditionally the Itô stochastic integral against dw_t is introduced in a different way, with discussion of which we start the chapter.

1. The classical definition

Let $(\Omega, \mathcal{F}, P)$ be a complete probability space, $\mathcal{F}_t$, $t \geq 0$, an increasing filtration of σ-fields $\mathcal{F}_t \subset \mathcal{F}$, and $w_t, t \geq 0$, a Wiener process relative to $\mathcal{F}_t$.

1. Definition. Let $f_t = f_t(\omega)$ be a function defined on $\Omega \times [0, \infty)$. We write $f \in H_0$ if there exist nonrandom points $0 = t_0 \leq t_1 \leq ... \leq t_n < \infty$ such that the f_{t_i} are $\mathcal{F}_{t_i}$-measurable, $Ef_{t_i}^2 < \infty$, and $f_t = f_{t_i}$ for $t \in [t_i, t_{i+1})$ if $i \leq n$, whereas $f_t = 0$ for $t \geq t_n$.

2. Exercise. Why does it not make much sense to consider functions satisfying $f_t = f_{t_i}$ for $t \in (t_i, t_{i+1}]$?

For $f \in H_0$ we set

$$If = \sum_{i=0}^{n-1} (w_{t_{i+1}} - w_{t_i}) f_{t_i}.$$

Obviously this definition is independent of the partition $\{t_i\}$ of $[0,\infty)$ provided that $f \in H_0$. In particular, the notation If makes sense, and I is a linear operator on H_0.

3. Lemma. *If $f \in H_0$, then*

$$E(If)^2 = E\int_0^\infty f_t^2\,dt, \qquad EIf = 0.$$

Proof. We have (see Theorem 3.1.12)

$$Ef_{t_j}^2(w_{t_{j+1}} - w_{t_j})^2 = Ef_{t_j}^2 E\{(w_{t_{j+1}} - w_{t_j})^2 | \mathcal{F}_{t_j}\} = Ef_{t_j}^2(t_{j+1} - t_j),$$

since $w_{t_{j+1}} - w_{t_j}$ is independent of $\mathcal{F}_{t_j}$ and f_{t_j} is $\mathcal{F}_{t_j}$-measurable. This and Cauchy's inequality imply that the first expression in the following relations makes sense:

$$Ef_{t_i}(w_{t_{i+1}} - w_{t_i})f_{t_j}(w_{t_{j+1}} - w_{t_j})$$
$$= Ef_{t_i}(w_{t_{i+1}} - w_{t_i})f_{t_j}E\{(w_{t_{j+1}} - w_{t_j})|\mathcal{F}_{t_j}\} = 0$$

if $i < j$, since $t_{i+1} \le t_j$ and $f_{t_j}, w_{t_{i+1}} - w_{t_i}, f_{t_i}$ are $\mathcal{F}_{t_j}$-measurable, whereas $w_{t_{j+1}} - w_{t_j}$ is independent of $\mathcal{F}_{t_j}$. Hence

$$E(If)^2 = \sum_{j=0}^{n-1} Ef_{t_j}^2(w_{t_{j+1}} - w_{t_j})^2 + 2\sum_{i<j\le n-1} Ef_{t_i}(w_{t_{i+1}} - w_{t_i})f_{t_j}(w_{t_{j+1}} - w_{t_j})$$
$$= \sum_{j=0}^{n-1} Ef_{t_j}^2(t_{j+1} - t_j) = E\int_0^\infty f_t^2\,dt.$$

Similarly, $Ef_{t_j}(w_{t_{j+1}} - w_{t_j}) = 0$ and $EIf = 0$. The lemma is proved.

The next step was not done in Secs. 2.7 and 2.8 because we did not have the necessary tools at that time. In the following lemma we use the notion of continuous time martingale, which is introduced in the same way as in Definition 3.2.1, just allowing m and n to be arbitrary numbers satisfying $0 \le n \le m$.

4. Lemma. *For $f \in H_0$, define $I_s f = I(I_{[0,s)}f)$. Then $(I_s f, \mathcal{F}_s)$ is a martingale for $s \ge 0$.*

Proof. Fix s and without loss of generality assume that $s \in \{t_0, ..., t_n\}$. If $s = t_k$, then

$$I_{[0,s)}f_t = \sum_{i=0}^{k-1} f_{t_i} I_{[t_i,t_{i+1})}(t), \quad I_s f = \sum_{i=0}^{k-1} f_{t_i}(w_{t_{i+1}} - w_{t_i}).$$

It follows that $I_s f$ is $\mathcal{F}_s$-measurable. Furthermore, if $t \leq s$, $t, s \in \{t_0, ..., t_n\}$, $t = t_r$, and $s = t_k$, then

$$E\{I_s f - I_t f | \mathcal{F}_t\} = \sum_{i=r}^{k-1} E\{f_{t_i} E\{w_{t_{i+1}} - w_{t_i} | \mathcal{F}_{t_i}\} | \mathcal{F}_t\} = 0.$$

The lemma is proved.

Next by using the theory of martingales we derive an inequality allowing us to define stochastic integrals with variable upper limit as continuous processes.

5. Lemma (Doob's inequality). *For $f \in H_0$, we have*

$$E \sup_{s \geq 0} (I_s f)^2 \leq 4E \int_0^\infty f_t^2 \, dt. \tag{1}$$

Proof. First, notice that

$$I_s f = \sum_{i=0}^{n-1} f_{t_i} (w_{t_{i+1} \wedge s} - w_{t_i \wedge s}).$$

Therefore, the process $I_s f$ is continuous in s and the sup in (1) can be taken over the set of rational s. In particular, the sup is a random variable, and (1) makes sense.

Next, let $0 \leq s_0 \leq ... \leq s_m < \infty$. Since $(I_{s_k} f, \mathcal{F}_{s_k})$ is a martingale, by Doob's inequality

$$E \sup_{k \leq m} (I_{s_k} f)^2 \leq 4E(I_{s_m} f)^2 = 4E \int_0^{s_m} f_t^2 \, dt \leq 4E \int_0^\infty f_t^2 \, dt.$$

Clearly the inequality between the extreme terms holds for any $s_0, ..., s_m$, not necessarily ordered. In particular, one can number all rationals on $[0, \infty)$ and then take the first $m + 1$ rational numbers as $s_0, ..., s_m$. If after that one lets $m \to \infty$, then one gets (1) by the monotone convergence theorem. The lemma is proved.

Lemma 3 allows us to follow an already familiar pattern. Namely, consider H_0 as a subset of $L_2(\mathcal{F} \otimes \mathfrak{B}[0, \infty), \mu)$, where $\mu(d\omega dt) = P(d\omega)dt$. On H_0 we have defined the operator I which maps H_0 isometrically to a subset of $L_2(\mathcal{F}, P)$. By Lemma 2.3.12 the operator I admits a unique extension to an isometric operator acting from $\bar{H}_0$ into $L_2(\mathcal{F}, P)$. We keep the same notation I for the extension, and for a function $f \in \bar{H}_0$ we define its *Itô stochastic integral* by the formula

$$\int_0^\infty f_t\,dw_t = If.$$

We have to explain that this integral coincides with the one introduced in Sec. 2.7.

6. Remark. Obviously

$$H_0 \subset H,$$

where H is introduced in Definition 2.8.1 as the set of all real-valued $\mathcal{F}_t$-adapted functions $f_t(\omega)$ which are $\mathcal{F} \otimes \mathfrak{B}(0,\infty)$-measurable and belong to $L_2(\mathcal{F} \otimes \mathfrak{B}[0,\infty),\mu)$.

7. Remark. Generally the processes from H_0 are not predictable, since they are right continuous in t. However, if one redefines them at points of discontinuity by taking the left limits, then one gets left-continuous, hence predictable, processes (see Exercise 2.8.3) coinciding with the initial ones for almost all t. It follows that $H_0 \subset L_2(\mathcal{P},\mu)$.

Observe that $\bar{H}_0$, which is the closure of H_0 in $L_2(\mathcal{F} \otimes \mathfrak{B}[0,\infty),\mu)$, coincides with the closure of H_0 in $L_2(\mathcal{P},\mu)$, since $L_2(\mathcal{P},\mu) \subset L_2(\mathcal{F}\otimes\mathfrak{B}[0,\infty),\mu)$. Furthermore, replacing the left continuous $\rho_n(t)$ in the proof of Theorem 2.8.2 with the right continuous $2^{-n}[2^n t]$, we see that $f \in \bar{H}_0$ if f_t is an $\mathcal{F}_t$-adapted $\mathcal{F}\otimes\mathfrak{B}(0,\infty)$-measurable function belonging to $L_2(\mathcal{F}\otimes\mathfrak{B}(0,\infty),\mu)$. In other words,

$$H \subset \bar{H}_0.$$

8. Remark. It follows by Theorem 2.8.8 (i) that If coincides with the Itô stochastic integral introduced in Sec. 2.7 on functions $f \in H_0$. Since $H_0 \subset H$ and H_0 is dense in H (Remarks 6 and 7) and $H = L_2(\mathcal{P},\mu)$ in the sense described in Exercise 2.8.5, we have that $\bar{H}_0 = L_2(\mathcal{P},\mu)$, implying that both stochastic integrals are defined on the same set and coincide there.

9. Definition. For $f \in \bar{H}_0$ and $s \ge 0$ define

$$\int_0^s f_t\,dw_t = \int_0^\infty I_{[0,s)}(t) f_t\,dw_t.$$

This is the traditional way of introducing the stochastic Itô integral against the Wiener process with variable upper limit. Notice that for many other martingales such as $m_t := \pi_t - t$, where π_t is a Poisson process with parameter one, it is much more natural to replace $I_{[0,s)}$ with $I_{[0,s]}$, since then $\int_0^s 1\,dm_t = m_s$ on each trajectory. In our situation the integral on each particular trajectory makes no sense, and taking $I_{[0,s]}$ leads to the same result since $I_{[0,s]} = I_{[0,s)}$ as elements of $L_2(\mathcal{F} \otimes \mathfrak{B}(0,\infty),\mu)$.

Defining the stochastic integral as result of a mapping into $L_2(\mathcal{F}, P)$ specifies the result only almost surely, so that for any $s \geq 0$ there are many candidates for $\int_0^s f_t\, dw_t$. If one chooses these candidates arbitrary for each s, one can easily end up with a process which has nonmeasurable trajectories for each ω. It is very important for the theory of stochastic integration that one can arrange the choosing in such a way that almost all trajectories become continuous in s.

10. Theorem. *Let $f \in \bar{H}_0$. Then the process $\int_0^s f_t\, dw_t$ admits a continuous modification.*

Proof. Take $f^n \in H_0$ such that

$$E \int_0^\infty |f_t - f_t^n|^2\, dt \leq 2^{-n}.$$

Then for each $s \geq 0$ in the sense of convergence in $L_2(\mathcal{F} \otimes \mathfrak{B}(0,\infty), \mu)$ we have

$$I_{[0,s)}(t) f_t = I_{[0,s)}(t) f_t^1 + I_{[0,s)}(t)(f_t^2 - f_t^1) + ... + I_{[0,s)}(t)(f_t^{n+1} - f_t^n) + ...$$

Hence by continuity (or isometry), in the sense of the mean-square convergence we have

$$\int_0^s f_t\, dw_t = \int_0^s f_t^1\, dw_t + \int_0^s (f_t^2 - f_t^1)\, dw_t + ... + \int_0^s (f_t^{n+1} - f_t^n)\, dw_t + ... \tag{2}$$

Here each term is continuous as the integral of an H_0-function, so that to prove the theorem it suffices to prove that the series in (2) converges uniformly for almost every ω.

By Doob's inequality

$$E \sup_{s \geq 0} \Big| \int_0^s (f_t^{n+1} - f_t^n)\, dw_t \Big|^2 \leq 4E \int_0^\infty (f_t^{n+1} - f_t^n)^2\, dt \leq 16 \cdot 2^{-n}.$$

By Chebyshev's inequality

$$P\{\sup_{s \geq 0} \Big| \int_0^s (f_t^{n+1} - f_t^n)\, dw_t \Big| \geq n^{-2}\} \leq 16 n^4 2^{-n},$$

and since the series $\sum n^4 2^{-n}$ converges, by the Borel-Cantelli lemma with probability one for all large n we have

$$\sup_{s\geq 0}\Big|\int_0^s (f_t^{n+1}-f_t^n)\,dw_t\Big|<n^{-2}.$$

Finally, we remember that $\sum n^{-2}<\infty$. The theorem is proved.

2. Properties of the stochastic integral on H

The Itô integral is defined on the set $\bar{H}_0$, which is a space of equivalence classes. By Remarks 1.7 and 1.8, in each equivalence class there is a function belonging to H. As usual we prefer to deal not with equivalence classes but rather with their particular representatives, and now we concentrate on integrating processes of class H. Furthermore, Theorem 1.10 allows us to consider only continuous versions of stochastic integrals.

1. Theorem. *Let $f,g\in H$, $a,b\in\mathbb{R}$. Then:*

(i) (*linearity*) *for all t at once with probability one*

$$\int_0^t (af_s+bg_s)\,dw_s=a\int_0^t f_s\,dw_s+b\int_0^t g_s\,dw_s; \tag{1}$$

(ii) $E\int_0^\infty f_t\,dw_t=0$;

(iii) *the process $\int_0^t f_s\,dw_s$ is a martingale relative to $\mathcal{F}_t^P$;*

(iv) *Doob's inequality holds:*

$$E\sup_t\Big(\int_0^t f_s\,dw_s\Big)^2\leq 4E\int_0^\infty f_t^2\,dt;$$

(v) *if $A\in\mathcal{F}$, $T\in[0,\infty]$, and $f_t(\omega)=g_t(\omega)$ for all $\omega\in A$ and $t\in[0,T)$, then*

$$I_A\int_0^t f_s\,dw_s=I_A\int_0^t g_s\,dw_s \tag{2}$$

for all $t\in[0,T]$ at once with probability one.

Proof. (i) For each $t\geq 0$ equation (1) (a.s.) follows by definition (see Lemma 2.3.12). Furthermore, both sides of (1) are continuous in t, and hence they coincide for all t if they coincide for all *rational* t. Since for each rational t, (1) holds almost surely and the set of rational numbers is countable and the intersection of countably many events of full probability has probability one, (1) indeed holds for all $t\geq 0$ on a set of full probability.

(ii) Take $f^n \in H_0$ such that $f^n \to f$ in $L_2(\mathcal{F} \otimes \mathfrak{B}(0,\infty), \mu)$. Then use Cauchy's inequality and Lemma 1.3 to find that

$$|EIf| = |EI(f - f^n)| \leq \big[EI(f - f^n)^2\big]^{1/2} = \big[E\int_0^\infty (f_t - f_t^n)^2\,dt\big]^{1/2} \to 0.$$

(iii) Take the same sequence f^n as above and remember that Lemma 1.4 allows us to write

$$E\{\int_0^t f_s^n\,dw_s|\mathcal{F}_r\} = \int_0^r f_s^n\,dw_s \quad \text{(a.s.)} \quad \forall 0 \leq r \leq t. \tag{3}$$

Furthermore, $E\big(\int_0^t f_s^n\,dw_s - \int_0^t f_s\,dw_s\big)^2 = E\int_0^t (f_s^n - f_s)^2\,ds \to 0$, and evaluating conditional expectation is a continuous operator in $L_2(\mathcal{F}, P)$ as a projection operator in $L_2(\mathcal{F}, P)$ (Theorem 3.1.14). Hence upon passing to the limit in (3) in the mean-square sense, we get an equality which shows that

(a) $\int_0^r f_s\,dw_s$ is $\mathcal{F}_r^P$-measurable as a function almost surely equal to an $\mathcal{F}_r$-measurable $E(\cdot|\mathcal{F}_r)$;

(b) the martingale equality holds.

This proves (iii). Assertion (iv) is proved in the same way as Lemma 1.5.

(v) By the argument in (i) it suffices to prove that (2) holds with probability one for each particular $t \in [0, T]$. In addition, $I_t f = I(I_{[0,t)}f)$, which shows that it only remains to prove that

$$I_A \int_0^\infty f_s\,dw_s = I_A \int_0^\infty g_s\,dw_s$$

(a.s.) if $f_s(\omega) = g_s(\omega)$ for all $\omega \in A$ and $s \geq 0$. But this is just statement (ii) of Theorem 2.8.8. The theorem is proved.

Further properties of the stochastic integral are related to the notion of stopping time (Definition 2.5.7), which, in particular, will allow us to extend the domain of definition of the stochastic integral from H to a larger set.

2. Exercise*. Prove that if a random process ξ_t is right continuous for each ω (or left continuous for each ω), then it is $\mathcal{F} \otimes \mathfrak{B}[0,\infty)$-measurable.

3. Exercise*. Let τ be an $\mathcal{F}_t$-stopping time. Prove that $I_{t<\tau}$ and $I_{t\leq\tau}$ are $\mathcal{F}_t$-adapted and $\mathcal{F} \otimes \mathfrak{B}[0,\infty)$-measurable, and that $\{\omega : t \leq \tau\} \in \mathcal{F}_t$ for every $t \geq 0$.

4. Exercise*. Let ξ_t be an $\mathcal{F}_t$-adapted continuous real-valued process, and take real numbers $a < b$. Define $\tau = \inf\{t \geq 0 : \xi_t \notin (a,b)\}$ $(\inf \emptyset := \infty)$ so that τ is the first exit time of ξ_t from (a,b). Prove that τ is an $\mathcal{F}_t$-stopping time.

The major part of stopping times we are going to deal with will be particular applications of Exercise 4 and the following.

5. Lemma. *Let $f = f_t(\omega)$ be nonnegative, $\mathcal{F}_t$-adapted, and $\mathcal{F} \otimes \mathfrak{B}[0,\infty)$-measurable. Assume that the σ-fields $\mathcal{F}_t$ are complete, that is, $\mathcal{F}_t = \mathcal{F}_t^P$. Then, for any $t \geq 0$, $\int_0^t f_s\,ds$ is $\mathcal{F}_t$-measurable.*

Proof. If the assertion holds for $f \wedge n$ in place of f, then by letting $n \to \infty$ and using the monotone convergence theorem we get the result for our f. It follows that without losing generality we may assume that f is bounded. Furthermore, we can cut off the function f in t by taking $t \geq 0$ and setting $f_s = 0$ for $s \geq t$. Then we see that it suffices to prove our assertion for $f \in H$.

In that case, as in the proof of Theorem 2.8.2 we conclude that there exist $f^n \in H_0$ such that

$$E\Big|\int_0^t f_s\,ds - \int_0^t f_s^n\,ds\Big| \leq E\int_0^t |f_s - f_s^n|\,ds \leq \sqrt{t}\Big(E\int_0^\infty |f_s - f_s^n|^2\,ds\Big)^{1/2} \to 0.$$

Furthermore, $\int_0^t f_s^n\,ds$ is obviously written as a sum in which all terms are $\mathcal{F}_t$-measurable. The mean-square limit of $\mathcal{F}_t$-measurable variables is at least $\mathcal{F}_t^P$-measurable, and the lemma is proved.

6. Remark. Due to this lemma, *everywhere below we assume that $\mathcal{F}_t^P = \mathcal{F}_t$*. This assumption does not restrict generality at all, since, as is easy to see, $(w_t, \mathcal{F}_t^P)$ is again a Wiener process and passing from $\mathcal{F}_t$ to $\mathcal{F}_t^P$ only enlarges the set H. Actually, the change of H is almost unnoticeable, since the set $\bar{H}_0$ remains unchanged as well as the stochastic integral and the inclusions $H_0 \subset H \subset \bar{H}_0$ hold true as before. Also, as we have pointed out before, we always take continuous versions of stochastic integrals, which is possible due to Theorem 1.10.

Before starting to use stopping times we point out two standard ways of approximating an arbitrary stopping time τ with discrete ones τ_n. One can use (2.5.3), or alternatively one lets

$$\tau_n(\omega) = (k+1)2^{-n} \quad \text{if} \quad \tau(\omega) \in [k2^{-n}, (k+1)2^{-n})$$

and $\tau_n(\omega) = \infty$ if $\tau(\omega) = \infty$. In other words,

$$\tau_n = 2^{-n}[2^n\tau] + 2^{-n}. \tag{4}$$

Then $\tau \le \tau_n$, $\tau_n - \tau \le 2^{-n}$, and

$$\{\omega : t < \tau_n(\omega)\} = \{\omega : 2^{-n}[2^n t] \le \tau\} \in \mathcal{F}_{2^{-n}[2^n t]} \subset \mathcal{F}_t,$$

so that τ_n are stopping times.

7. Theorem. *Let $f \in H$. Denote $\xi_t = \int_0^t f_s\,dw_s$, $t \in [0,\infty]$, and let τ be an $\mathcal{F}_t$-stopping time. Then*

$$\xi_\tau = \int_0^\infty I_{s<\tau} f_s\,dw_s = \int_0^\infty I_{s\le\tau} f_s\,dw_s \tag{5}$$

(a.s.) and (Wald's identity)

$$E\Big(\int_0^\tau f_s\,dw_s\Big)^2 = E\int_0^\tau f_s^2\,ds. \tag{6}$$

Proof. To prove (5), first assume that τ takes only countably many values $\{t_1, t_2, ...\}$. On the set $\Omega_k := \{\omega : \tau(\omega) = t_k\}$ we have $I_{s<t_k} f_s = I_{s<\tau} f_s$ for all $s \ge 0$. By definition, on Ω_k (a.s.) we have

$$\xi_\tau = \xi_{t_k} = \int_0^\infty I_{s<t_k} f_s\,dw_s,$$

and by Theorem 1 (v) on Ω_k (a.s.)

$$\int_0^\infty I_{s<t_k} f_s\,dw_s = \int_0^\infty I_{s<\tau} f_s\,dw_s.$$

Thus the first equality in (5) holds on any Ω_k (a.s.). Since $\bigcup_k \Omega_k = \Omega$, this equality holds almost surely. To prove it in the general case it suffices to define τ_n by (4) and notice that $\xi_{\tau_n} \to \xi_\tau$ because ξ_t is continuous, whereas

$$E\Big(\int_0^\infty I_{s<\tau} f_s\,dw_s - \int_0^\infty I_{s<\tau_n} f_s\,dw_s\Big)^2 = E\int_0^\infty I_{\tau\le s<\tau_n} f_s^2\,ds \to 0$$

by the dominated convergence theorem. The second equality in (5) is obvious since the integrands coincide for almost all (ω, s).

On the basis of (5) and the isometry of stochastic integration we conclude that

$$E\xi_\tau^2 = E\int_0^\infty I_{s<\tau} f_s^2\,ds = E\int_0^\tau f_s^2\,ds.$$

The theorem is proved.

The following fundamental inequality can be extracted from the original memoir of Itô [**It**].

8. Theorem. *Let $f \in H$, and let $N, c > 0$, and $T \leq \infty$ be constants. Then*

$$P\{\sup_{t\leq T}\Big|\int_0^t f_s\,dw_s\Big| \geq c\} \leq P\{\int_0^T f_s^2\,ds \geq N\} + \frac{1}{c^2}E\big(N \wedge \int_0^T f_s^2\,ds\big).$$

Proof. We use the standard way of stopping stochastic integrals

$$\xi_t = \int_0^t f_s\,dw_s$$

by using their "brackets", defined as

$$\langle\xi\rangle_t := \int_0^t f_s^2\,ds.$$

Let $\tau = \inf\{t \geq 0 : \langle\xi\rangle_t \geq N\}$, so that τ is the first exit time of $\langle\xi\rangle_t$ from $(-1, N)$. By Exercise 4 and Lemma 5 we have that τ is a stopping time. Furthermore,

$$\{\omega : \tau < T\} \subset \{\omega : \langle\xi\rangle_T \geq N\}$$

and on the set $\{\omega : \tau \geq T\}$ we have $I_{s<\tau} f_s = f_s$ if $s < T$. Therefore, upon denoting

$$A = \{\omega : \sup_{t\leq T} |\xi_t| \geq c\},$$

by the Doob-Kolmogorov inequality for submartingales we get

$$P(A, \tau \geq T) = P\{\tau \geq T, \sup_{t\leq T}\Big|\int_0^t I_{s<\tau} f_s\,dw_s\Big| \geq c\}$$

$$\leq P\{\sup_{t\leq T}\Big|\int_0^t I_{s<\tau} f_s\,dw_s\Big|^2 \geq c^2\} \leq \frac{1}{c^2}E\Big(\int_0^T I_{s<\tau} f_s\,dw_s\Big)^2$$

$$= \frac{1}{c^2}E\int_0^{T\wedge\tau} I_{s<\tau} f_s^2\,ds = \frac{1}{c^2}E\Big(\int_0^T I_{s<\tau} f_s^2\,ds \wedge \int_0^\tau I_{s<\tau} f_s^2\,ds\Big)$$

$$\leq \frac{1}{c^2}E\big(N \wedge \int_0^T f_s^2\,ds\big),$$

where in the last inequality we have used the fact that, if $\tau<\infty$, then obviously $\langle\xi\rangle_\tau=N$, and if $\tau=\infty$, then $\langle\xi\rangle_\tau\le N$. Hence

$$P(A)=P(A,\tau<T)+P(A<\tau\ge T)\le P(\tau<T)+P(A,\tau\ge T)$$

$$\le P\{\int_0^T f_s^2\,ds\ge N\}+\frac{1}{c^2}E\big(N\wedge\int_0^T f_s^2\,ds\big).$$

The theorem is proved.

9. Exercise. Under the assumptions of Theorem 8, prove that

$$P(\langle\xi\rangle_T\ge N)\le\frac{1}{N}E\big(c^2\wedge\sup_{t\le T}\xi_t^2\big)+P(\sup_{t\le T}|\xi_t|\ge c).$$

10. Exercise. Prove Davis's inequality: If $f\in H$, then

$$\frac{1}{3}E\langle\xi\rangle_T^{1/2}\le E\sup_{t\le T}|\xi_t|\le 3E\langle\xi\rangle_T^{1/2}.$$

3. Defining the Itô integral if $\int_0^T f_s^2\,ds<\infty$

Denote by $\mathcal{S}$ the set of all $\mathcal{F}_t$-adapted, $\mathcal{F}\otimes\mathfrak{B}(0,\infty)$-measurable processes f_t such that

$$\int_0^T f_s^2\,ds<\infty\quad\text{(a.s.)}\quad\forall T<\infty.$$

Our task here is to define $\int_0^t f_t\,dw_t$ for $f\in\mathcal{S}$.

Define

$$\tau(n)=\inf\{t\ge 0:\int_0^t f_s^2\,ds\ge n\}.$$

In Sec. 2 we have already seen that $\tau(n)$ are stopping times and

$$\int_0^{\tau(n)} f_s^2\,ds\le n.$$

Furthermore, obviously $\tau(n)\uparrow\infty$ (a.s.) as $n\to\infty$. Finally, notice that $I_{s<\tau(n)}f_s\in H$. Indeed, the fact that this process is $\mathcal{F}_t$-adapted follows from Exercise 2.3. Also

$$E\int_0^\infty I_{s<\tau(n)}f_s^2\,ds=E\int_0^{\tau(n)} f_s^2\,ds\le n<\infty.$$

It follows from the above that the stochastic integrals

$$\xi_t(n) := \int_0^t I_{s<\tau(n)} f_s \, dw_s$$

are well defined. If $\int_0^t f_s \, dw_s$ were defined, it would certainly satisfy

$$\int_0^t I_{s<\tau(n)} f_s \, dw_s = \int_0^{t\wedge\tau(n)} f_s \, dw_s.$$

This observation is a clue to defining $\int_0^t f_s \, dw_s$.

1. Lemma. *Let $f \in \mathcal{S}$. Then there exists a set $\Omega' \subset \Omega$ such that $P(\Omega') = 1$ and, for every $\omega \in \Omega'$, $m \geq n$, and $t \in [0, \tau(n, \omega)]$, we have $\xi_t(n) = \xi_t(m)$.*

Proof. Fix t and n, and notice that on the set $A = \{\omega : t \leq \tau(n)\}$ we have

$$I_{s<t\wedge\tau(n)} f_s = I_{s<t\wedge\tau(m)} f_s$$

for all s. By Theorem 2.1, almost surely on A

$$\int_0^t I_{s<\tau(n)} f_s \, dw_s = \int_0^\infty I_{s<t\wedge\tau(n)} f_s \, dw_s = \int_0^t I_{s<\tau(m)} f_s \, dw_s.$$

In other words, almost surely

$$I_{t\leq\tau(n)} \int_0^t I_{s<\tau(n)} f_s \, dw_s = I_{t\leq\tau(n)} \int_0^t I_{s<\tau(m)} f_s \, dw_s \tag{1}$$

for any t and $m \geq n$. Clearly, the set Ω' of all ω for each of which (1) holds for all $m \geq n$ and rational t has full probability. If $\omega \in \Omega'$, then (1) is actually true for all t, since both sides are left continuous in t. This is just a restatement of our assertion, so the lemma is proved.

2. Corollary. *If $f \in \mathcal{S}$, then with probability one the sequence $\xi_t(n)$ converges uniformly on each finite time interval.*

3. Definition. Let $f \in \mathcal{S}$. For those ω for which the sequence $\xi_t(n)$ converges uniformly on each finite time interval we define

$$\int_0^t f_s \, dw_s = \lim_{n\to\infty} \int_0^t I_{s<\tau(n)} f_s \, dw_s.$$

For all other ω we define $\int_0^t f_s \, dw_s = 0$.

Of course, one has to check that Definition 3 does not lead to anything new if $f\in H$. Observe that if $f\in H$, then by Fatou's theorem and the dominated convergence theorem

$$E\Big|\lim_{n\to\infty}\int_0^t I_{s<\tau(n)}f_s dw_s-\int_0^t f_s dw_s\Big|^2\le\lim_{n\to\infty}E\int_0^t(1-I_{s<\tau(n)})f_s^2\,ds=0.$$

Therefore both definitions give the same result almost surely for any given t. Since Definition 3 yields a continuous process, we see that, for $f\in H$, the new integral is also the integral in the previous sense.

Also notice that $f\equiv 1\in\mathcal{S}$, $\tau(n)=n$, and hence (a.s.)

$$\int_0^t 1\cdot dw_s=\lim_{n\to\infty}\int_0^t I_{s<n}\,dw_s=\lim_{n\to\infty}\int_0^\infty I_{s<n\wedge t}\,dw_s=\lim_{n\to\infty}w_{n\wedge t}=w_t.$$

Now come some properties of the stochastic integral on $\mathcal{S}$.

4. Exercise. By using Fatou's theorem and Exercise 2.10, prove Davis's inequality for $f\in\mathcal{S}$.

5. Theorem. *Let $f,f^n,g\in\mathcal{S}$, and let $\delta,\varepsilon>0$, $T\in[0,\infty)$ be constants. Then:*

(i) *the stochastic integral $\int_0^t f_s\,dw_s$ is continuous in t and $\mathcal{F}_t$-adapted;*

(ii) *we have*

$$P\{\sup_{t\le T}\Big|\int_0^t f_s\,dw_s-\int_0^t g_s\,dw_s\Big|\ge\varepsilon\}\le P\{\int_0^T|f_s-g_s|^2\,ds\ge\delta\}$$

$$+\frac{1}{\varepsilon^2}E\delta\wedge\int_0^T(f_s-g_s)^2\,ds\le P\{\int_0^T|f_s-g_s|^2\,ds\ge\delta\}+\frac{\delta}{\varepsilon^2};\qquad(2)$$

(iii) *we have*

$$\int_0^T|f_s^n-f_s|^2\,ds\xrightarrow{P}0\Longrightarrow\sup_{t\le T}\Big|\int_0^t f_s^n\,dw_s-\int_0^t f_s\,dw_s\Big|\xrightarrow{P}0.$$

Proof. (i) The continuity of $\int_0^t f_s\,ds$ follows from Definition 3, in which

$$\int_0^t I_{s<\tau(n)}f_s\,ds$$

are continuous and $\mathcal{F}_t$-adapted (even $\mathcal{F}_t$-martingales). Their limit is also $\mathcal{F}_t$-adapted.

To prove (ii), first notice that all expressions in (2) are monotone and right continuous in ε and δ. Therefore, it suffices to prove (2) only at points of their continuity. Also notice that the second inequality in (2) is obvious since $\delta \wedge \cdot \leq \delta$.

Now fix appropriate ε, and δ and define

$$\tau(n) = \inf\{t \geq 0 : \int_0^t f_s^2\, ds \geq n\}, \quad \sigma(n) = \inf\{t \geq 0 : \int_0^t g_s^2\, ds \geq n\},$$

$$f_s^n = I_{s<\tau(n)} f_s, \quad g_s^n = I_{s<\sigma(n)} g_s.$$

Since f^n and g^n belong to H, inequality (2) holds with f^n, g^n in place of f, g due to the linearity of the stochastic integral on H and Theorem 2.8. Furthermore, almost surely, as $n \to \infty$,

$$\sup_{t \leq T} \Big| \int_0^t f_s^n\, dw_s - \int_0^t g_s^n\, dw_s \Big| \to \sup_{t \leq T} \Big| \int_0^t f_s\, dw_s - \int_0^t g_s\, dw_s \Big|,$$

$$\int_0^T |f_s^n - g_s^n|^2\, ds \to \int_0^T |f_s - g_s|^2\, ds.$$

These convergences of random variables imply convergence of the corresponding distribution functions at all points of their continuity. Adding to this tool the dominated convergence theorem, we get (2) from its version for f^n, g^n.

To prove (iii) it suffices to take $g = f^n$ in (2) and let first $n \to \infty$ and then $\delta \downarrow 0$. The theorem is proved.

6. Exercise. Prove that the converse implication in assertion (iii) of Theorem 5 is also true.

Before discussing further properties of the stochastic integral we denote

$$\chi_n(x) = (-n) \vee x \wedge n,$$

so that $\chi_n(x) = x$ for $|x| \leq n$ and $\chi_n(x) = n \operatorname{sign} x$ otherwise. Observe that, if $f \in \mathcal{S}$, $T \in [0, \infty)$, and $f_s^n := \chi_n(f_s) I_{s<T}$, then $f^n \in H$ and (a.s.)

$$\int_0^T |f_s^n - f_s|^2\, ds \to 0.$$

This way of approximating $f \in \mathcal{S}$ along with Theorem 5 and known properties of the stochastic integral on H immediately yields assertions (i) through (iii) of the following theorem.

7. Theorem. (i) *If* $f, g \in \mathcal{S}$, $a, b \in \mathbb{R}$, *then* (*a.s.*)

$$\int_0^t (af_s + bg_s)\,dw_s = a\int_0^t f_s\,dw_s + b\int_0^t g_s\,dw_s \quad \forall t \in [0,\infty).$$

(ii) *If* $f_s = f_{t_i}$ *for* $s \in [t_i, t_{i+1})$, $0 = t_0 < t_1 < ...$, $t_i \to \infty$ *as* $i \to \infty$, *and the* f_{t_i} *are* $\mathcal{F}_{t_i}$*-measurable, then* $f \in \mathcal{S}$ *and* (*a.s.*) *for every* $t \geq 0$

$$\int_0^t f_s\,dw_s = \sum_{t_{i+1}<t} f_{t_i}(w_{t_{i+1}} - w_{t_i}) + f_{t_k}(w_t - w_{t_k}),$$

where k *is such that* $t_k \leq t$ *and* $t_{k+1} > t$.

(iii) *If* $f, g \in \mathcal{S}$, $T < \infty$, $A \in \mathcal{F}$, *and* $f_s(\omega) = g_s(\omega)$ *for all* $s \in [0,T]$ *and* $\omega \in A$, *then almost surely on* A

$$\int_0^t f_s\,dw_s = \int_0^t g_s\,dw_s \quad \forall t \leq T.$$

(iv) *If* $f \in \mathcal{S}$, $T < \infty$, *and* τ *is a stopping time satisfying* $\tau \leq T$, *then* (*a.s.*)

$$\int_0^\tau f_s\,dw_s = \int_0^T I_{s<\tau} f_s\,dw_s. \tag{3}$$

Assertion (iv) is obtained from the fact that due to Theorem 2.7, if $f \in H$, then the left-hand side of (3) equals (a.s.)

$$\int_0^{\tau\wedge T} f_s\,dw_s = \int_0^\infty I_{s<\tau\wedge T} f_s\,dw_s = \int_0^\infty I_{s<T} I_{s<\tau} f_s\,dw_s = \int_0^T I_{s<\tau} f_s\,dw_s.$$

In the statement of the following theorem we use the fact that if $f \in \mathcal{S}$, τ is a stopping time, and

$$E\int_0^\tau f_s^2\,ds = E\int_0^\infty I_{s<\tau} f_s^2\,ds < \infty,$$

then $I_{s<\tau} f_s \in H$ and $\int_0^\infty I_{s<\tau} f_s\,dw_s$ makes sense.

8. Theorem. *Let* $f \in \mathcal{S}$, $T \leq \infty$, *and let* τ *be an almost surely finite stopping time* (*that is,* $\tau(\omega) < \infty$ *for almost every* ω).

(i) *If* $E\int_0^\tau f_s^2\,ds < \infty$, *then*

$$\int_0^{\tau} f_s\,dw_s = \int_0^{\infty} I_{s<\tau} f_s\,dw_s \quad (a.s.) \tag{4}$$

and Wald's identities hold:

$$E\Big(\int_0^{\tau} f_s\,dw_s\Big)^2 = E\int_o^{\tau} f_s^2\,ds, \quad E\int_0^{\tau} f_s\,dw_s = 0. \tag{5}$$

In particular (for $f \equiv 1$), if $E\tau < \infty$, then $Ew_\tau^2 = E\tau$ and $Ew_\tau = 0$.

(ii) *If $E\int_0^T f_t^2\,dt < \infty$, then $\int_0^t f_s\,dw_s$ is a martingale for $t \in [0,T] \cap [0,\infty)$.*

Proof. To prove (4) it suffices to remember what has been said before the theorem and use Theorems 7 and 2.7, which imply that (a.s.)

$$\int_0^{\tau} f_s\,dw_s = \lim_{n\to\infty}\int_0^{\tau\wedge n} f_s\,dw_s = \lim_{n\to\infty}\int_0^{n} I_{s<\tau\wedge n} f_s\,dw_s$$

$$= \lim_{n\to\infty}\int_0^{\infty} I_{s<n} I_{s<\tau\wedge n} f_s\,dw_s = \lim_{n\to\infty}\int_0^{\infty} I_{s<\tau\wedge n} f_s\,dw_s = \int_0^{\infty} I_{s<\tau} f_s\,dw_s,$$

where the last equality holds (a.s.) because

$$E\int_0^{\infty} |I_{s<\tau\wedge n} f_s - I_{s<\tau} f_s|^2\,ds \to 0.$$

Equation (5) follows from (4) and the properties of the stochastic integral on H.

To prove (ii) it suffices to notice that

$$\int_0^t f_s\,dw_s = \int_0^t I_{s<T} f_s\,dw_s \quad \text{for} \quad t \le T, \quad I_{s<T} f_s \in H,$$

and that stochastic integrals of elements of H are martingales. The theorem is proved.

9. Example. Sometimes Wald's identities can be used for concrete computations. To show an example, let $a, b > 0$, and let τ be the first exit time of w_t from $(-a, b)$. Then, for each t, we have $|w_{t\wedge\tau}| \le a + b$ and

$$(a+b)^2 \ge Ew_{t\wedge\tau}^2 = E\Big(\int_0^t I_{s<\tau}\,dw_s\Big)^2 = E\int_0^t I_{s<\tau}\,ds = Et\wedge\tau.$$

Since this is true for any t, by the monotone convergence theorem $E\tau \leq (a+b)^2 < \infty$.

It follows that Wald's identities hold true for this τ, so that, in particular, $Ew_\tau = 0$, which is written as

$$-aP(w_\tau = -a) + bP(w_\tau = b) = 0.$$

Adding to this that $P(w_\tau = -a) + P(w_\tau = b) = 1$ since $\tau < \infty$ (a.s.), we get

$$P(w_\tau = -a) = \frac{b}{a+b}, \quad P(w_\tau = b) = \frac{a}{a+b}.$$

Furthermore, $Ew_\tau^2 = E\tau$. In other words,

$$E\tau = a^2 P(w_\tau = -a) + b^2 P(w_\tau = b) = ab.$$

We thus rediscover the results of Exercise 2.6.5.

10. Remark. Generally Wald's identities are wrong if $E\int_0^\tau f_s^2\,ds = \infty$. For instance, let $\tau = \inf\{t \geq 0 : w_t \geq 1\}$. We know that τ has Wald's distribution, so that $P(\tau < \infty) = 1$ and $w_\tau = 1$ (a.s.). Hence $Ew_\tau = 1 \neq 0$, and one of identities is violated. It follows that $E\tau = \infty$ and $Ew_\tau^2 = 1 \neq E\tau$.

11. Exercise. In Remark 10 we have that $E\sqrt{\tau} = \infty$, which follows from the explicit formula for Wald's distribution. In connection with this, prove that, if $E(\int_0^\tau f_s^2\,ds)^{1/2} < \infty$, then $E\int_0^\tau f_s\,dw_s = 0$.

Regarding assertion (ii) of Theorem 8, it is worth noting that generally stochastic integrals are not martingales. We give two exercises to that effect: Exercises 12 and 7.4.

12. Exercise. For $t < 1$ consider the process $1 + \int_0^t (1-s)^{-1}\,dw_s$ and let τ be the first time it hits zero. Prove that $\tau < 1$ (a.s.) and

$$\int_0^t \frac{1}{1-s} I_{s<\tau}\,dw_s + 1 = 0$$

(a.s.) for all $t \geq 1$.

13. Exercise. Prove that if $E\big(\int_0^T f_s^2\,ds\big)^{1/2} < \infty$, then $\int_0^t f_s\,dw_s$ is a martingale for $t \leq T$.

In the future we also use the stochastic integral with variable lower limit. If $0 \leq t_1 \leq t_2 < \infty$, $f_t(\omega)$ is measurable with respect to (ω, t) and $\mathcal{F}_t$-adapted, and

$$\int_{t_1}^{t_2} f_t^2\,dt < \infty \quad \text{(a.s.)},$$

then define

$$\int_{t_1}^{t_2} f_s\,dw_s = \int_0^{t_2} I_{[t_1,t_2)}(s) f_s\,dw_s. \tag{6}$$

We have $I_{[t_1,t_2)} = I_{[0,t_2)} - I_{[0,t_1)}$. Hence, if $f \in \mathcal{S}$, then (a.s.)

$$\int_{t_1}^{t_2} f_s\,dw_s = \int_0^{t_2} f_s\,dw_s - \int_0^{t_1} f_s\,dw_s. \tag{7}$$

14. Theorem. *Let $f \in \mathcal{S}$, $0 \le t_1 \le t_2 < \infty$, and let g be $\mathcal{F}_{t_1}$-measurable. Then (a.s.)*

$$g \int_{t_1}^{t_2} f_s\,dw_s = \int_{t_1}^{t_2} g f_s\,dw_s, \tag{8}$$

that is, one can factor out appropriately measurable random variables.

Proof. First of all, notice that the right-hand side of (8) is well defined by virtue of definition (6) but not (7). Next, (8) is trivial for $f \in H_0$, since both sides are just simple sums. If $f \in H$, one can approximate f with $f^n \in H_0$ so that

$$\int_0^\infty |f_t^n - f_t|^2\,dt \xrightarrow{P} 0, \quad \int_0^\infty |g f_t^n - g f_t|^2\,dt \xrightarrow{P} 0.$$

Then one can pass to the limit on the basis of Theorem 5. After having proved (8) for $f \in H$, one easily gets (8) in the general case by noticing that in the very Definition 3 we use $f^n \in H$ such that $\int_0^T |f_t^n - f_t|^2\,dt \to 0$ (a.s.) for every $T < \infty$. The theorem is proved.

4. Itô integral with respect to a multidimensional Wiener process

1. Definition. Let $(\Omega, \mathcal{F}, P)$ be a complete probability space, let $\mathcal{F}_t, t \ge 0$, be a filtration of *complete* σ-fields $\mathcal{F}_t \subset \mathcal{F}$, and let $w_t = (w_t^1, ..., w_t^d)$ be a d-dimensional process on Ω. We say that w_t is a d-dimensional Wiener process relative to $\mathcal{F}_t, t \ge 0$, or that $(w_t, \mathcal{F}_t)$ is a d-dimensional Wiener process, if

(i) w_t^k are Wiener processes for each $k = 1, ..., d$,

(ii) the processes $w_t^1, ..., w_t^d$ are independent,

(iii) w_t is $\mathcal{F}_t$-adapted and $w_{t+h} - w_t$ is independent of $\mathcal{F}_t$ if $t, h \geq 0$.

If $f_t = (f_t^1, ..., f_t^d)$ is a d-dimensional process, we write $f \in \mathcal{S}$ whenever $f^i \in \mathcal{S}$ for any i. If $f_t = (f_t^1, ..., f_t^d) \in \mathcal{S}$, we define

$$\int_0^t f_s \, dw_s = \int_0^t f_s^1 \, dw_s^1 + ... + \int_0^t f_s^d \, dw_s^d, \tag{1}$$

so that $f_s \, dw_s$ is interpreted as the scalar product of f_s and dw_s.

The stochastic integral against multidimensional Wiener processes possesses properties quite similar to the ones in the one-dimensional case. We neither list nor prove all of them, pointing out only that, if $f, g \in \mathcal{S}$, $T < \infty$, and $E \int_0^T (|f_s|^2 + |g_s|^2) \, ds < \infty$, then

$$E \int_0^T f_s \, dw_s \int_0^T g_s \, dw_s = E \int_0^T f_s \cdot g_s \, ds.$$

This property is easily proved on the basis of (1) and the fact that, for instance,

$$E \int_0^T f_s^1 \, dw_s^1 \int_0^T g_s^2 \, dw_s^2 = 0,$$

which in turn is almost obvious for $f, g \in H_0$ and extends to $f, g \in \mathcal{S}$ by standard passages to the limit.

We also need to integrate matrix-valued processes. If $\sigma_t = (\sigma_t^{ik}), i = 1, ..., d_1$, $k = 1, ..., d$, and $\sigma^{ij} \in \mathcal{S}$, then we write $\sigma \in \mathcal{S}$ and by

$$\int_0^t \sigma_s \, dw_s$$

we naturally mean the d_1-dimensional process, the ith coordinate of which is given by

$$\sum_{k=1}^d \int_0^t \sigma_s^{ik} \, dw_s^k.$$

In other terms we look at $\sigma_s \, dw_s$ as the product of the matrix σ_s and the column vector dw_s.

2. Exercise*. Prove that if $E \int_0^t \operatorname{tr} \sigma_s \sigma_s^* \, ds < \infty$, then

$$E\Big| \int_0^t \sigma_s \, dw_s \Big|^2 = E \int_0^t \operatorname{tr} \sigma_s \sigma_s^* \, ds.$$

3. Exercise. Let b_t be a d-dimensional process, $b_t \in \mathcal{S}$. Prove that

$$\Big(\exp\Big(\int_0^t b_t \, dw_t - (1/2) \int_0^t |b_t|^2 \, dt\Big), \mathcal{F}_t \Big)$$

is a supermartingale.

5. Itô's formula

In the usual calculus, after the notion of integral is introduced one discusses the rules of integration and compiles the table of "elementary" integrals. The most important tools of integration are change of variable and integration by parts, which are proved on the basis of the formula for differentiating superpositions. The formula for the *stochastic* differential of a superposition is called *Itô's formula*. This formula was discovered in [**It**] as a curious fact and then became the main tool of modern stochastic calculus.

1. Definition. Let $(\Omega, \mathcal{F}, P)$ be a complete probability space carrying a d_1-dimensional Wiener process $(w_t, \mathcal{F}_t)$ and a continuous d-dimensional $\mathcal{F}_t$-adapted process ξ_t. Assume that we are also given a $d \times d_1$ matrix valued process σ_t and a d-dimensional process b_t such that $\sigma \in \mathcal{S}$ and b is jointly measurable in (ω, t), $\mathcal{F}_t$-adapted, and $\int_0^T |b_s| \, ds < \infty$ (a.s.) for any $T < \infty$. Then we write

$$d\xi_t = \sigma_t \, dw_t + b_t \, dt$$

if and only if (a.s.) for all t

$$\xi_t = \xi_0 + \int_0^t \sigma_s \, dw_s + \int_0^t b_s \, ds. \tag{1}$$

In that case one says that ξ_t has *stochastic differential* equal to $\sigma_t \, dw_t + b_t \, dt$.

From calculus we know that if $f(x)$ and $g(t)$ are differentiable, then

$$df(g(t)) = f'(g(t)) \, dg(t).$$

It turns out that stochastic differentials possess absolutely different properties. For instance, consider $d(w_t^2)$ for one-dimensional w_t. If the usual rules were true, we would have $dw_t^2 = 2w_t \, dw_t$, that is,

$$w_t^2 = 2 \int_0^t w_s \, dw_s.$$

However, this is impossible since

$$Ew_t^2 = t, \quad E\int_0^t w_s^2\,ds < \infty, \quad E\int_0^t w_s\,dw_s = 0.$$

Still, there is a case in which the usual formula holds. This case was found by Hitsuda. Let (w'_t, w''_t) be a two-dimensional Wiener process and define the *complex* Wiener process by

$$z_t = w'_t + iw''_t.$$

It turns out (see Exercise 5) that, for any analytic function $f(z)$, we have $df(z_t) = f'(z_t)\,dz_t$, that is,

$$f(z_t) = f(0) + \int_0^t f(z_s)\,dz_s. \tag{2}$$

We have what would be "the usual formula" if z_t were piecewise differentiable.

We have introduced formal d_1-dimensional expressions $\sigma_t\,dw_t + b_t\,dt$. Now we define rules of operating with them. We assume that while multiplying them by constants, adding up, and evaluating their scalar products the usual algebraic rules of factoring out and combining similar terms are enforced *along with* the following multiplication table (which, by the way, keeps the products of stochastic differentials in the set of stochastic differentials):

$$dw_t^i dw_t^j = \delta^{ij}\,dt, \quad dw_t^i dt = (dt)^2 = 0. \tag{3}$$

A crucial role in the proof of Itô's formula is played by the following.

2. Lemma. *Let ξ_t, η_t be real-valued processes having stochastic differentials. Then $\xi_t\eta_t$ also has a stochastic differential, and*

$$d(\xi_t\eta_t) = \eta_t\,d\xi_t + \xi_t\,d\eta_t + (d\xi_t)d\eta_t.$$

Proof. Let

$$\xi_t = \xi_0 + \int_0^t \sigma_s\,dw_s + \int_0^t b_s\,ds, \quad \eta_t = \eta_0 + \int_0^t \tilde\sigma_s\,dw_s + \int_0^t \tilde b_s\,ds,$$

where σ_s and $\tilde\sigma_s$ are vector-valued processes and b_s and $\tilde b_s$ are real-valued ones. By the above rules, assuming the summation convention, we can write

$$\eta_t d\xi_t = \eta_t(\sigma_t^k \, dw_t^k + b_t \, dt) = \eta_t \sigma_t^k \, dw_t^k + \eta_t b_t \, dt = \eta_t \sigma_t \, dw_t + \eta_t b_t \, dt,$$

$$\xi_t d\eta_t = \xi_t \tilde{\sigma}_t \, dw_t + \xi_t \tilde{b}_t \, dt, \quad (d\xi_t) d\eta_t = \sigma_t^j \, dw_t^j \tilde{\sigma}_t^k \, dw_t^k = \sigma_t^j \tilde{\sigma}_t^j \, dt = \sigma_t \cdot \tilde{\sigma}_t \, dt.$$

Therefore our assertion means that, for all $t \in [0, \infty)$ at once, with probability one,

$$\xi_t \eta_t = \xi_0 \eta_0 + \int_0^t (\eta_s \sigma_s + \xi_s \tilde{\sigma}_s) \, dw_s + \int_0^t (\eta_s b_s + \xi_s \tilde{b}_s + \sigma_s \cdot \tilde{\sigma}_s) \, ds. \tag{4}$$

First, notice that the right-hand side of (4) makes sense because (a.s.)

$$\int_0^t |\eta_s b_s| \, ds \le \max_{s \le t} |\eta_s| \int_0^t |b_s| \, ds < \infty,$$

$$\int_0^t |\eta_s \sigma_s^j|^2 \, ds \le \max_{s \le t} |\eta_s| \int_0^t |\sigma_s^j|^2 \, ds < \infty,$$

$$\int_0^t |\sigma_s \cdot \tilde{\sigma}_s| \, ds \le \int_0^t |\sigma_s|^2 \, ds + \int_0^t |\tilde{\sigma}_s|^2 \, ds < \infty.$$

Next, notice that if $d\xi_t' = \sigma_t' \, dw_t + b_t' \, dt$ and $d\xi_t'' = \sigma_t'' \, dw_t + b_t'' \, dt$ and (4) holds with ξ', σ', b' and ξ'', σ'', b'' in place of ξ, σ, b, then it also holds for $\xi' + \xi'', \sigma' + \sigma'', b' + b''$. It follows that we may concentrate only on two possibilities for $d\xi_t$: $d\xi_t = \sigma_t \, dw_t$ and $d\xi_t = b_t \, dt$. We have the absolutely similar situation with η. Therefore, we have to deal only with four pairs of $d\xi_t$ and $d\eta_t$. To finish our preparation, we also notice that both sides of (4) are continuous in t, so that to prove that they coincide with probability one for all t at once, it suffices to prove that they are equal almost surely for each particular t.

Thus, fix t, and first let $d\xi_t = b_t \, dt$ and $d\eta_t = \tilde{b}_t \, dt$. Then (4) follows from the usual calculus (or is proved as in the following case).

The two cases, (i) $d\xi_t = \sigma_t \, dw_t$ and $d\eta_t = \tilde{b}_t \, dt$ and (ii) $d\xi_t = b_t \, dt$ and $d\eta_t = \tilde{\sigma}_t \, dw_t$, are similar, and we concentrate on (i).

Let $0 = t_{m0} \le t_{m1} \le \ldots \le t_{mk_m} = t$ be a sequence of partitions of $[0, t]$ such that $\max_i (t_{m,i+1} - t_{mi}) \to 0$ as $m \to \infty$. Define

$$\kappa_m(s) = t_{mi}, \quad \tilde{\kappa}_m(s) = t_{m,i+1} \quad \text{if} \quad s \in [t_{mi}, t_{m,i+1}).$$

Obviously $\kappa_m(s), \tilde{\kappa}_m(s) \to s$ uniformly on $[0, t]$. In addition, the formula

$$ab - cd = (a - c)d + (b - d)a$$

and Theorem 3.14 show that (a.s.)

$$\xi_t\eta_t - \xi_0\eta_0 = \sum_{i=0}^{k_m-1} (\xi_{t_{m,i+1}}\eta_{t_{m,i+1}} - \xi_{t_{mi}}\eta_{t_{mi}})$$

$$= \sum_{i=0}^{k_m-1} \eta_{t_{mi}} \int_{t_{mi}}^{t_{m,i+1}} \sigma_s\,dw_s + \sum_{i=0}^{k_m-1} \xi_{t_{m,i+1}} \int_{t_{mi}}^{t_{m,i+1}} \tilde{b}_s\,ds$$

$$= \int_0^t \eta_{\kappa_m(s)}\sigma_s\,dw_s + \int_0^t \xi_{\tilde{\kappa}_m(s)}\tilde{b}_s\,ds. \tag{5}$$

Furthermore, as $m \to \infty$, we have (a.s.)

$$\Big|\int_0^t \xi_{\tilde{\kappa}_m(s)}\tilde{b}_s\,ds - \int_0^t \xi_s\tilde{b}_s\,ds\Big| \le \sup_{s\le t}|\xi_{\tilde{\kappa}_m(s)} - \xi_s| \int_0^t |\tilde{b}_s|\,ds \to 0,$$

$$\int_0^t |\eta_{\kappa_m(s)} - \eta_s|^2(\sigma_s^j)^2\,ds \le \sup_{s\le t}|\eta_{\kappa_m(s)} - \eta_s|^2 \int_0^t (\sigma_s^j)^2\,ds \to 0,$$

and the last relation by Theorem 3.5 (iii) implies that

$$\int_0^t \eta_{\kappa_m(s)}\sigma_s\,dw_s \xrightarrow{P} \int_0^t \eta_s\sigma_s\,dw_s. \tag{6}$$

Now by letting $m \to \infty$ in (5) we get (4) (a.s.) in our particular case.

Thus it only remains to consider the case $d\xi_t = \sigma_t\,dw_t$, $d\eta_t = \tilde{\sigma}_t\,dw_t$, and prove that

$$\xi_t\eta_t = \xi_0\eta_0 + \int_0^t (\eta_s\sigma_s + \xi_s\tilde{\sigma}_s)\,dw_s + \int_0^t \sigma_s\cdot\tilde{\sigma}_s\,ds. \tag{7}$$

Notice that we may assume that $\xi_0 = \eta_0 = 0$, since in the initial reduction to four cases we could absorb the initial values in the terms with dt.

Now we again use bilinearity and conclude that, since σ and $\tilde{\sigma}$ can be represented as sums of vector-valued processes each of which has only one nonidentically zero element, we only have to prove (7) for such simple vector-valued processes. Furthermore, keeping in mind that each $f \in \mathcal{S}$ can be approximated by $f^n \in H_0$ (see, for instance, the proof of Theorem 3.14), we see that we may assume that $\sigma^j, \tilde{\sigma}^j \in H_0$.

In this way we conclude that to prove (7) in the general case, it suffices to prove that, if $f, g \in H_0$, $\xi_r = \int_0^r f_s\,dw_s^i$, and $\eta_r = \int_0^r g_s\,dw_s^j$, then (a.s.)

$$\xi_t\eta_t = \int_0^t f_s\eta_s\,dw_s^i + \int_0^t g_s\xi_s\,dw_s^j + \int_0^t f_s g_s\delta^{ij}\,ds. \tag{8}$$

Remember that t is fixed, and without losing generality assume that the partitions corresponding to f and g coincide and t is one of the partition points. Let $\{t_0, t_1, ...\}$ be the common partition with $t = t_k$. Next, as above we take the sequence of partitions defined by t_{mi} of $[0, t]$ and again without loss of generality assume that each t_i lying in $[0, t]$ belongs to $\{t_{mi} : i = 0, 1, ...\}$. We use the formula

$$ab - cd = (a - c)d + (b - d)c + (a - c)(b - d) \tag{9}$$

and Theorem 3.14. Fix a $q = 0, ..., k - 1$ and, by default summing up with respect to those r for which $t_q \leq t_{mr} < t_{q+1}$, write (a.s.)

$$\xi_{t_{q+1}}\eta_{t_{q+1}} - \xi_{t_q}\eta_{t_q} = \sum(\xi_{t_{m,r+1}}\eta_{t_{m,r+1}} - \xi_{t_{mr}}\eta_{t_{mr}})$$

$$= \sum \eta_{t_{mr}} \int_{t_{mr}}^{t_{m,r+1}} f_s\,dw_s^i + \sum \xi_{t_{mr}} \int_{t_{mr}}^{t_{m,r+1}} g_s\,dw_s^j$$

$$+ \sum \int_{t_{mr}}^{t_{m,r+1}} f_s\,dw_s^i \int_{t_{mr}}^{t_{m,r+1}} g_s\,dw_s^j = \int_{t_q}^{t_{q+1}} \eta_{\kappa_m(s)} f_s\,dw_s^i +$$

$$+ \int_{t_q}^{t_{q+1}} \xi_{\kappa_m(s)} g_s\,dw_s^j + f_{t_q}g_{t_q}\sum(w_{t_{m,r+1}}^i - w_{t_{mr}}^i)(w_{t_{m,r+1}}^j - w_{t_{mr}}^j). \tag{10}$$

In the expression after the last equality sign the first two terms converge in probability to

$$\int_{t_q}^{t_{q+1}} \eta_s f_s\,dw_s^i, \quad \int_{t_q}^{t_{q+1}} \xi_s g_s\,dw_s^j$$

respectively, which is proved in the same way as (6). If $i = j$, the last term converges in probability to

$$f_{t_q}g_{t_q}(t_{q+1} - t_q) = \int_{t_q}^{t_{q+1}} f_s g_s\,ds$$

by Theorem 2.2.6. Consequently, by letting $m \to \infty$ in (10) and then adding up the results for $q = 0, ..., k - 1$, we come to (7) if $i = j$. For $i \neq j$ one uses the same argument complemented by the observation that the last sum in (10) tends to zero in probability, since its mean is zero due to the independence of w^i and w^j, and

$$E\Big[\sum(w^i_{t_{m,r+1}} - w^i_{t_{mr}})(w^j_{t_{m,r+1}} - w^j_{t_{mr}})\Big]^2 = \operatorname{Var}\big[...\big]$$

$$= \sum E(w^i_{t_{m,r+1}} - w^i_{t_{mr}})^2(w^j_{t_{m,r+1}} - w^j_{t_{mr}})^2$$

$$= \sum(t_{m,r+1} - t_{mr})^2 \leq \max_i(t_{m,i+1} - t_{mi})t \to 0.$$

The lemma is proved.

3. Exercise. Explain why in the treatment of the fourth case one cannot use a formula similar to (5) in place of (10).

4. Theorem (Itô's formula). *Let a d_1-dimensional process ξ_t have stochastic differential, and let $u(x) = u(x^1, ..., x^{d_1})$ be a real-valued twice continuously differentiable function of $x \in \mathbb{R}^{d_1}$. Then $u(\xi_t)$ has a stochastic differential, and*

$$du(\xi_t) = u_{x^i}(\xi_t)\, d\xi^i_t + (1/2)u_{x^ix^j}(\xi_t)\, d\xi^i_t d\xi^j_t. \tag{11}$$

Proof. Let C^2 be the set of all real-valued twice continuously differentiable function on $\mathbb{R}^{d_1}$. We are going to use the fact that for every $u \in C^2$ there is a sequence of polynomials u^m such that $u^m, u^m_{x^i}, u^m_{x^ix^j}$ converge to $u, u_{x^i}, u_{x^ix^j}$ uniformly on each ball. For such a sequence and any ω, t, i, j

$$\sup_{s\leq t}|u^m_{x^i}(\xi_t) - u_{x^i}(\xi_t)| + \sup_{s\leq t}|u^m_{x^ix^j}(\xi_t) - u_{x^ix^j}(\xi_t)| \to 0,$$

since each trajectory of ξ_s, $s \leq t$, lies in a ball. It follows easily that, if (11) is true for u^m, then it is also true for u.

Thus, we only need to prove (11) for polynomials, and to do this it obviously suffices to show that (11) holds for linear function and also for the product of any two functions u and v for each of which (11) holds.

For linear u formula (11) is obvious. If (11) holds for u and v, then by Lemma 2

$$d(u(\xi_t)v(\xi_t)) = u(\xi_t)\, dv(\xi_t) + v(\xi_t)\, du(\xi_t) + (du(\xi_t))dv(\xi_t)$$

$$= [uv_{x^i} + vu_{x^i}](\xi_t)\, d\xi^i_t + (1/2)[uv_{x^ix^j} + vu_{x^ix^j}](\xi_t)\, d\xi^i_t d\xi^i_t + u_{x^i}v_{x^j}(\xi_t)\, d\xi^i_t d\xi^i_t$$

$$= (uv)_{x^i}(\xi_t)\, d\xi^i_t + (1/2)(uv)_{x^ix^j}(\xi_t)\, d\xi^i_t d\xi^i_t.$$

The theorem is proved.

Itô's formula (11) looks very much like Taylor's formula with two terms. Usually one rewrites it in a different way. Namely, let $d\xi_t = \sigma_t\, dw_t + b_t\, dt$, $a = (1/2)\sigma_t\sigma_t^*$. Simple manipulations show that $(d\xi_t^i)d\xi_t^j = 2a_t^{ij}\, dt$ and hence

$$du(\xi_t) = L_t u(\xi_t)\, dt + \sigma_t^* u_x(\xi_t)\, dw_t,$$

where $u_x = \operatorname{grad} u$ is a column vector and L_t is the second-order differential operator given by

$$L_t v(x) = a_t^{ij} v_{x^i x^j}(x) + b_t^i v_{x^i}(x).$$

In this notation (11) means that for all t (a.s.)

$$u(\xi_t) = u(\xi_0) + \int_0^t L_s u(\xi_s)\, ds + \int_0^t \sigma_s^* u_x(\xi_s)\, dw_s. \tag{12}$$

5. Exercise. Prove that (2) holds for analytic functions f.

Itô's formula leads to extremely important formulas relating the theory of stochastic integration with the theory of partial differential equations. One of them is the following theorem.

6. Theorem. *Let ξ_0 be nonrandom, let Q be a domain in $\mathbb{R}^{d_1}$, let $\xi_0 \in Q$, let τ be the first exit time of ξ_t from Q, and let u be a function which is continuous in $\bar{Q}$ and has continuous first and second derivatives in Q. Assume that*

$$P(\tau < \infty) = 1, \quad E\int_0^\tau |L_s u(\xi_s)|\, ds < \infty.$$

Then

$$u(\xi_0) = Eu(\xi_\tau) - E\int_0^\tau L_s u(\xi_s)\, ds.$$

We give no proof to this theorem because it is just a particular result, and usually when one needs such results it is easier and shorter to prove what is needed directly instead of trying to find the corresponding result in the literature. We will see examples of this in Sec. 7.

Roughly speaking, to prove Theorem 6 one plugs τ in place of t in (12) and takes expectations. The main difficulties on the way are caused by the fact that u is not even given in the whole $\mathbb{R}^{d_1}$ and the expectation of a stochastic integral does not necessarily exist, let alone equal zero. One overcomes these difficulties by taking smaller domains $Q_m \uparrow Q$, extending u outside Q_m, taking τ even smaller that the first exit time from Q_m, and then passing to the limit.

6. An alternative proof of Itô's formula

The approach we have in mind is based on using stopping times and stochastic intervals. It turns out that these tools could be used right from the beginning, even for defining Itô integral. First we briefly outline how to do this, to give the reader one more chance to go through the basics of the theory and also to show a way which is valid for integrals against more general martingales.

1. Definition. Let $\tau = \tau(\omega)$ be a $[0, \infty)$-valued function on Ω taking only finitely many values, say $t_1, ..., t_n \geq 0$. We say that τ is a *simple stopping time* (relative to $\mathcal{F}_t$) if $\{\omega : \tau(\omega) = t_k\} \in \mathcal{F}_{t_k}$ for any $k = 1, ..., n$. The set of all simple stopping times is denoted by $\mathcal{M}$.

Below in this section we only use simple stopping times.

2. Exercise*. (i) Prove that simple stopping times are stopping times, and that $\{\omega : \tau(\omega) \geq t\} \in \mathcal{F}_t$ for any t.

(ii) Derive from (i) that if τ_1 and τ_2 are simple stopping times, then $\tau_1 \wedge \tau_2$ and $\tau_1 \vee \tau_2$ are simple stopping times as well.

3. Lemma. *For a real-valued function $\gamma(\omega)$, define the* stochastic interval *$(\!(0, \gamma]\!]$ as the set $\{(\omega, t) : \omega \in \Omega, 0 < t \leq \gamma(\omega)\}$ and let Π be the collection of all stochastic intervals $(\!(0, \tau]\!]$ with τ running through the set of all simple stopping times. Finally, for $\Delta = (\!(0, \tau]\!] \in \Pi$, define $\zeta(\Delta) = w_\tau$. Then ζ is a random orthogonal measure on Π with reference measure $\mu = P \times \ell$ and $E\zeta(\Delta) = 0$ for any $\Delta \in \Pi$.*

Proof. Let τ be a simple stopping time and $\{t_1, ..., t_n\}$ the set of its values. Then

$$w_\tau = w_{t_1} I_{\tau = t_1} + ... + w_{t_n} I_{\tau = t_n}$$

and, since $Ew_s^2 < \infty$, $E\zeta^2((\!(0, \tau]\!]) = Ew_\tau^2 < \infty$.

Next we will be using the simple fact that, if τ is a simple stopping time and the set $\{0 = t_0 < t_1 < ... < t_n\}$ contains all possible values of τ, then

$$w_\tau = \sum_{i=0}^{n-1} f_{t_i}(w_{t_{i+1}} - w_{t_i}), \quad \tau = \sum_{i=0}^{n-1} f_{t_i}(t_{i+1} - t_i), \tag{1}$$

where $f_t := I_{\tau > t}$ is $\mathcal{F}_t$-measurable (Exercise 2). Since $\{\omega : \tau(\omega) > t_i\} \in \mathcal{F}_{t_i}$ and $w_{t_{i+1}} - w_{t_i}$ is independent of $\mathcal{F}_{t_i}$, we have

$$Ef_{t_i}(w_{t_{i+1}} - w_{t_i}) = Ef_{t_i}E(w_{t_{i+1}} - w_{t_i}) = 0, \quad E\zeta((\!(0, \tau]\!]) = 0.$$

Now, let τ and σ be simple stopping times, $\{t_1, ..., t_n\}$ the ordered set of their values, and $\Delta_1 = (0, \tau]\!]$ and $\Delta_2 = (0, \sigma]\!]$. By using (1) we have

$$E\zeta(\Delta_1)\zeta(\Delta_2) = Ew_\tau w_\sigma = \sum_{i,j=0}^{n-1} Ef_{t_i} g_{t_j}(w_{t_{i+1}} - w_{t_i})(w_{t_{j+1}} - w_{t_j}),$$

which, in the same way as in the proofs of Theorem 2.7.3 or Lemma 1.3 used in other approaches, is shown to be equal to

$$\sum_{i=0}^{n-1} Ef_{t_i} g_{t_i}(t_{i+1} - t_i) = E\sum_{i=0}^{n-1} I_{\tau\wedge\sigma > t_i}(t_{i+1} - t_i) = E\tau \wedge \sigma.$$

Since

$$E\tau \wedge \sigma = \int_\Omega \int_0^\infty I_{(0,\tau]\!]\cap(0,\sigma]\!]}(\omega, t)\, P(d\omega)dt = \mu(\Delta_1 \cap \Delta_2),$$

the lemma is proved.

From this lemma we derive the following version of Wald's identities.

4. Corollary. *Let τ_1 and τ_2 be simple stopping times. Then $Ew^2_{\tau_1} = E\tau_1$ and $E(w_{\tau_1} - w_{\tau_2})^2 = E|\tau_1 - \tau_2|$.*

Indeed, we get the first equality from the proof of Lemma 3 by taking $\sigma = \tau$. To prove the second one, define $\tau = \tau_1 \vee \tau_2$, $\sigma = \tau_1 \wedge \tau_2$ and notice that

$$E(w_{\tau_1} - w_{\tau_2})^2 = E(w_\tau - w_\sigma)^2 = Ew^2_\tau - 2Ew_\tau w_\sigma + Ew^2_\sigma$$
$$= E\tau - E\sigma = E(\tau - \sigma) = E|\tau_1 - \tau_2|.$$

5. Exercise. Carry over the result of Corollary 4 to all bounded stopping times.

6. Remark. Lemma 3 and the general Theorem 2.3.13 imply that there is a stochastic integral operator, say I, defined on $L_2(\Pi, \mu)$ with values in $L_2(\mathcal{F}, P)$. Since Π is a π-system of subsets of $\Omega \times (0, \infty)$, we have $L_2(\Pi, \mu) = L_2(\sigma(\Pi), \mu)$ due to Theorem 2.3.19.

7. Remark. It turns out that $\sigma(\Pi) = \mathcal{P}$. Indeed, on the one hand the indicators of the sets $(0, \tau]\!]$ generating $\sigma(\Pi)$ are left-continuous and $\mathcal{F}_t$-adapted, hence predictable (Exercise 2.8.3). In other words, $(0, \tau]\!] \in \mathcal{P}$ and $\sigma(\Pi) \subset \mathcal{P}$. On the other hand, if $A \in \mathcal{F}_s$, $s \geq 0$, and for $n > s$ we define $\tau_n = s$ on A and $\tau_n = n$ on $\Omega \setminus A$, then τ_n are simple stopping times and

$$(0, \tau_n]\!] = \{(\omega, t) : 0 < t \leq \tau_n(\omega)\}$$

$$= \{(\omega, t) : 0 < t \le s, \omega \in A\} \bigcup \{(\omega, t) : 0 < t \le n, \omega \in A^c\},$$

$$\bigcup_n (\!(0, \tau_n]\!] = (A \times (0, s]) \cup (A^c \times (0, \infty)) \in \sigma(\Pi),$$

so that $(\bigcup_n (\!(0, \tau_n]\!])^c = A \times (s, \infty) \in \sigma(\Pi)$. It follows that the set generating $\mathcal{P}$ is a subset of $\sigma(\Pi)$ and $\mathcal{P} \subset \sigma(\Pi)$.

8. Remark. Remark 7 and the definition of $L_2(\Pi, \mu)$ imply the somewhat unexpected result that for every $f \in L_2(\mathcal{P}, \mu)$, in particular, $f \in H$, there are simple stopping times τ_i^m and *constants* c_i^m defined for $m = 1, 2, ...$ and $i = 1, ..., k(m) < \infty$ such that

$$E \int_0^\infty |f_t - \sum_{i=1}^{k(m)} c_i^m I_{(\!(0, \tau_i^m]\!]}(t)|^2 \, dt \to 0$$

as $m \to \infty$.

9. Exercise. Find simple stopping times τ_i^m and constants c_i^m such that, for the one-dimensional Wiener process w_t,

$$E \int_0^\infty |I_{t \le 1} w_t - \sum_{i=1}^{k(m)} c_i^m I_{(\!(0, \tau_i^m]\!]}(t)|^2 \, dt \to 0$$

as $m \to \infty$.

10. Remark. The operator I from Remark 6 coincides on $L_2(\Pi, \mu)$ with the operator of stochastic integration introduced before Remark 1.6. This follows easily from the uniqueness of continuation and Theorem 2.7, showing that the old stochastic integral coincides with the new one on the indicators of $(\!(0, \tau]\!]$ and both are equal to w_τ.

After making sure that we deal with the same objects as in Sec. 5, we start proving Itô's formula, allowing ourselves to use everything proved before Sec. 5. As in Sec. 5, we need only prove Lemma 5.2. Define $\kappa_n(t) = 2^{-n}[2^n t]$.

Due to (5.9) we have

$$w_t^i w_t^j = \int_0^t w_{\kappa_n(s)}^i \, dw_s^j + \int_0^t w_{\kappa_n(s)}^j \, dw_s^i$$

$$+ \sum_{k=0}^\infty (w^i_{\frac{k+1}{2^n} \wedge t} - w^i_{\frac{k}{2^n} \wedge t})(w^j_{\frac{k+1}{2^n} \wedge t} - w^j_{\frac{k}{2^n} \wedge t}), \quad i, j = 1, ..., d \quad \text{(a.s.)}. \qquad (2)$$

By sending n to infinity, from the theorem on quadratic variation of the Wiener process we get that (a.s.)

$$w^i_t w^j_t = \int_0^t w^i_s\, dw^j_s + \int_0^t w^j_s\, dw^i_s + \delta^{ij} t, \quad i,j = 1,...,d. \tag{3}$$

Furthermore, for $\gamma, \tau \in \mathcal{M}$, $\gamma \leq \tau$, by using the fact that the sets of all values of γ, τ are finite, we obtain that

$$\int_0^\infty w^j_\gamma I_{\gamma < s \leq \tau}\, dw^i_s = w^j_\gamma (w^i_\tau - w^i_\gamma) \quad \text{(a.s.)}.$$

Hence and from (3) for $i, j = 1, ..., d$, $\tau, \sigma \in \mathcal{M}$, $\gamma = \tau \wedge \sigma$ we have (a.s.)

$$w^i_\tau w^j_\sigma = (w^i_\tau - w^i_\gamma) w^j_\gamma + (w^j_\sigma - w^j_\gamma) w^i_\gamma + w^i_\gamma w^j_\gamma$$

$$= \int_0^\infty w^j_\gamma I_{\gamma < s \leq \tau}\, dw^i_s + \int_0^\infty w^i_\gamma I_{\gamma < s \leq \sigma}\, dw^j_s$$

$$+ \int_0^\infty w^j_s I_{s \leq \gamma}\, dw^i_s + \int_0^\infty w^i_s I_{s \leq \gamma}\, dw^j_s + \delta^{ij} \gamma$$

$$= \int_0^\infty w^j_{s\wedge\sigma} I_{s \leq \tau}\, dw^i_s + \int_0^\infty w^i_{s\wedge\tau} I_{s \leq \sigma}\, dw^j_s + \int_0^\infty I_{s \leq \tau} I_{s \leq \sigma}\, ds.$$

By replacing here τ, σ by $\tau \wedge t, \sigma \wedge t$, we conclude that (a.s.)

$$w^i_{t\wedge\tau} = \int_0^t I_{s \leq \tau}\, dw^i_s, \quad w^j_{t\wedge\sigma} = \int_0^t I_{s \leq \sigma}\, dw^j_s,$$

$$w^i_{t\wedge\tau} w^j_{t\wedge\sigma} = \int_0^t w^j_{s\wedge\sigma} I_{s \leq \tau}\, dw^i_s + \int_0^t w^i_{s\wedge\tau} I_{s \leq \sigma}\, dw^j_s + \int_0^t I_{s \leq \tau} I_{s \leq \sigma}\, ds. \tag{4}$$

Next, similarly to our argument about (2) and (3), by replacing w^j_t with t and then w^i_t with t as well, instead of (4) we get

$$t \wedge \sigma = \int_0^t I_{s \leq \sigma}\, ds, \quad t \wedge \tau = \int_0^t I_{s \leq \tau}\, ds,$$

$$
\begin{aligned}
(t \wedge \sigma) w^i_{t\wedge\tau} &= \int_0^t (s \wedge \sigma) I_{s\le\tau}\, dw^i_s + \int_0^t w^i_{s\wedge\tau} I_{s\le\sigma}\, ds, \\
(t \wedge \tau)(t \wedge \sigma) &= \int_0^t (s \wedge \sigma) I_{s\le\tau}\, ds + \int_0^t (s \wedge \tau) I_{s\le\sigma}\, ds.
\end{aligned} \tag{5}
$$

To finish the preliminaries, we observe that for each $\mathcal{F}_0$-measurable random variable ξ_0, obviously

$$
\xi_0 w^i_{t\wedge\tau} = \int_0^t \xi_0 I_{s\le\tau}\, dw^i_s, \quad (t\wedge\tau)\xi_0 = \int_0^t \xi_0 I_{s\le\tau}\, ds. \tag{6}
$$

Now we recall the notion of stochastic differential from before Lemma 5.2, and the multiplication table (5.3). Then we automatically have the following.

11. Lemma. *All the formulas (4), (5), and (6) can be written in one and the same way: If ξ_t, η_t are real-valued processes and*

$$
d\xi_t = \sigma_t\, dw_t + b_t\, dt, \quad d\xi_t = \sigma'_t\, dw_t + b'_t\, dt, \tag{7}
$$

where all entries of σ_t, σ'_t and of b_t, b'_t are indicators of elements of Π, then

$$
d(\xi_t \eta_t) = \xi_t\, d\eta_t + \eta_t\, d\xi_t + (d\xi_t)(d\eta_t). \tag{8}
$$

Also notice that since both sides of equality (8) are linear in ξ and in η, equality (8) immediately extends to all processes ξ_t, η_t satisfying (7) with functions σ, σ', b, b' of class $S(\Pi)$.

Now we are ready to prove Lemma 5.2, saying that (8) holds true for all scalar processes ξ_t, η_t possessing stochastic differentials. To this end, assume first that $\sigma', b' \in S(\Pi)$ and take a sequence of processes σ_n, b_n of class $S(\Pi)$ such that (a.s.)

$$
\int_0^T (|\sigma_t - \sigma_{nt}|^2 + |b_t - b_{nt}|)\, dt \to 0 \quad \forall T \in [0, \infty).
$$

Define also processes ξ^n_t, replacing σ, b in (6) by σ_n, b_n. As is well known, in probability

$$\sup_{t\le T}[|\int_0^t (\sigma_s-\sigma_{ns})\,dw_s|+|\int_0^t (b_s-b_{ns})\,ds|]\to 0,$$

$$\sup_{s\le T}|\xi_t-\xi^n_t|\to 0\quad \forall T\in[0,\infty). \tag{9}$$

If necessary, we take a subsequence and we assume that the convergences in (9) hold almost surely. Then by the dominated convergence theorem we have (a.s.)

$$\int_0^T |\xi_t-\xi^n_t|(|\sigma'_t|^2+|b'_t|)\,dt\to 0,$$

$$\int_0^T |\eta_t|(|\sigma_t-\sigma_{nt}|^2+|b_t-b_{nt}|)\,dt\to 0,$$

$$\int_0^T |\sigma_t\cdot\sigma'_t-\sigma_{nt}\cdot\sigma'_t|\,dt$$

$$\le (\int_0^T |\sigma_t-\sigma_{nt}|^2\,dt)^{1/2}(\int_0^T |\sigma'_t|^2\,dt)^{1/2}\to 0\quad \forall T\in[0,\infty).$$

This and an argument similar to the one which led us to (9) show that in the integral form of (8), with ξ^n_t instead of ξ_t, we can pass to the limit in probability and get (8) for the limit process ξ_t. Of course, after this we fix the process ξ_t and we carry out a similar limit passage in (8) affecting the second factor. In this way we get Lemma 5.2 in a straightforward way from the quite elementary Lemma 11.

7. Examples of applying Itô's formula

In this section w_t is a d-dimensional Wiener process.

1. Example. Let τ be the first exit time of w_t from $B_R=\{x:|x|<R\}$, where $R>0$ is a number. As we know, τ is a stopping time. Take

$$u(x)=(1/d)(R^2-|x|^2)$$

and apply Itô's formula to $u(w_t)$. Here $\xi_t=w_t$, σ is the identity matrix, $b=0$, and the corresponding differential operator $L_t=(1/2)\Delta$. We have (a.s.)

$$u(w_t)=-t-\int_0^t (2/d)w_s\,dw_s+(1/d)R^2\quad \forall t.$$

Substitute $t\wedge\tau$ in place of t, take expectations, and notice that, since $|w_t|\le R$ before τ, we have $0\le u(w_{t\wedge\tau})\le(1/d)R^2$ and

$$E\int_0^{t\wedge\tau}|w_s|^2\,ds\le R^2t<\infty.$$

Then we obtain

$$Eu(w_{t\wedge\tau})=-E(t\wedge\tau)+(1/d)R^2,\quad E(t\wedge\tau)=(1/d)R^2-Eu(w_{t\wedge\tau}).$$

It follows in particular that

$$E(t\wedge\tau)\le(1/d)R^2,\quad E\tau\le(1/d)R^2,\quad \tau<\infty\quad\text{(a.s.)}.$$

Furthermore, by letting $t\to\infty$ and noticing that on the set $\{\tau<\infty\}$ we obviously have $u(w_{t\wedge\tau})\to u(w_\tau)=0$, by the monotone convergence and dominated convergence theorems we conclude that

$$E\tau=\lim_{t\to\infty}E(t\wedge\tau)=(1/d)R^2-\lim_{t\to\infty}Eu(w_{t\wedge\tau})=(1/d)R^2.$$

Notice that, for $d=1$, we have the result which we know already: $E\tau=R^2$. Also notice that if we wanted to use Theorem 5.6, then we would have to find a function u such that $L_tu(w_t)=-1$ for $s\le\tau$ and $u(w_\tau)=0$. In other words, we needed u such that $(1/2)\Delta u=-1$ in B_R and $u=0$ on ∂B_R. This is exactly the one we used above. Finally, notice that in order to apply Theorem 5.6 we have to be sure in advance that $P(\tau<\infty)=1$.

2. Example. Fix $\varepsilon>0$ and $x_0\in\mathbb{R}^d$ with $|x_0|>\varepsilon$. Let us find the probability P that w_t will ever reach $\bar B_\varepsilon(x_0)=\{x:|x-x_0|\le\varepsilon\}$.

First find the probability P_R that w_t reaches $\{|x-x_0|=\varepsilon\}$ before reaching $\{|x-x_0|=R\}$, where $R>|x_0|$. We want to apply Theorem 5.6 and therefore represent the desired probability P_R as $E\phi(w_\tau)$, where τ is the first exit time of w_t from $\{x:\varepsilon<|x-x_0|<R\}$, $\phi=1$ on $\{|x-x_0|=\varepsilon\}$ and $\phi=0$ on $\{|x-x_0|=R\}$. Notice that, owing to Example 1, we have $\tau<\infty$ (a.s.).

Now it is natural to try to find a function u such that $u=\phi$ on $\{|x-x_0|=\varepsilon\}\cup\{|x-x_0|=R\}$ and $\Delta u=0$ in $\{x:\varepsilon<|x-x_0|<R\}$. This is natural, since then $P_R=u(0)$ by Theorem 5.6. It turns out that an appropriate function u exists and is given by

$$u(x)=\begin{cases}A(|x-x_0|^{-(d-2)}-R^{-(d-2)}) & \text{if}\quad d\ge3,\\ A(\ln|x-x_0|-\ln R) & \text{if}\quad d=2,\\ A(|x-x_0|-R) & \text{if}\quad d=1,\end{cases}$$

where

$$A = \begin{cases} (\varepsilon^{-(d-2)} - R^{-(d-2)})^{-1} & \text{if } \ d \geq 3, \\ (\ln \varepsilon - \ln R)^{-1} & \text{if } \ d = 2, \\ (\varepsilon - R)^{-1} & \text{if } \ d = 1. \end{cases}$$

Next, since the trajectories of w are continuous and for any T, ω are bounded on $[0, T]$, the event that w_t ever reaches $\bar{B}_\varepsilon(x_0)$ is the union of nested events, say E_n that w_t reaches $\{|x - x_0| = \varepsilon\}$ before reaching $\{|x - x_0| = n\}$. Hence $P = \lim_{n\to\infty} P_n$ and

$$P = \begin{cases} \frac{\varepsilon^{d-2}}{|x_0|^{d-2}} & \text{if } \ d \geq 3, \\ 1 & \text{if } \ d \leq 2. \end{cases}$$

We see that one- and two-dimensional Wiener processes reach any neighborhood of any point with probability one. For $d \geq 3$ this probability is strictly less than one, and this leads to the conjecture that $|w_t| \to \infty$ as $t \to \infty$ for $d \geq 3$.

3. Example. Our last example is aimed at proving the conjecture from the end of Example 2. Fix $x_0 \neq 0$ and take $\varepsilon > 0$ such that $\varepsilon < |x_0|$. Denote by τ_ε the first time w_t reaches $\{x : |x - x_0| \leq \varepsilon\}$.

First we prove that

$$\xi_t := |w_{t\wedge\tau_\varepsilon} - x_0|^{2-d}$$

is a bounded martingale. The boundedness of ξ_t is obvious: $0 < \xi_t \leq \varepsilon^{2-d}$. To prove that it is a martingale, construct a smooth function $f(x)$ on $\mathbb{R}^d$ such that $f(x) = |x - x_0|^{2-d}$ for $|x - x_0| \geq \varepsilon$. Then $\xi_t = f(w_{t\wedge\tau_\varepsilon})$, and by Itô's formula

$$f(w_t) = f(0) + \int_0^t (1/2)\Delta f(w_s)\,ds + \int_0^t f_x(w_s)\,dw_s.$$

Hence, owing to $\Delta f(x) = 0$, which holds for $|x - x_0| \geq \varepsilon$, we have

$$\xi_t = f(w_{t\wedge\tau_\varepsilon}) = |x_0|^{2-d} + \int_0^t I_{s\leq\tau_\varepsilon} f_x(w_s)\,dw_s.$$

Here the second term and the right-hand side are martingales since $|f_x|$ is bounded. By the theorem on convergence of nonnegative (super)martingales, $\lim_{t\to\infty} \xi_t$ exists with probability one. We certainly have to remember that this theorem was proved only for discrete time supermartingales. But its proof is based on Doob's upcrossing inequality, and for continuous supermartingales this inequality and the convergence theorem are extended without any difficulty as in the case of Lemma 1.5.

Now use that ξ_t is bounded to conclude that

$$|x_0|^{2-d} = E\xi_0 = E\xi_t = E\lim_{t\to\infty}\xi_t$$

$$= E\frac{1}{\lim_{t\to\infty}|w_t - x_0|^{d-2}} I_{\tau_\varepsilon=\infty} + EI_{\tau_\varepsilon<\infty}\varepsilon^{2-d}.$$

By using the result of Example 2 we get that the last expectation is $|x_0|^{2-d}$, and therefore

$$E\frac{1}{\lim_{t\to\infty}|w_t - x_0|^{d-2}} I_{\tau_\varepsilon=\infty} = 0,$$

so that $\lim_{t\to\infty}|w_t| = \infty$ (a.s.) on the set $\{\tau_\varepsilon = \infty\}$ for each $\varepsilon > 0$. Finally,

$$P\{\bigcup_{\varepsilon=1/m}\{\tau_\varepsilon = \infty\}\} = \lim_{m\to\infty} P(\tau_{1/m} = \infty) = \lim_{m\to\infty}(1 - 1/|mx_0|^{d-2}) = 1,$$

and $\lim_{t\to\infty}|w_t| = \infty$ (a.s.) indeed.

4. Exercise. Let $d = 2$ and take τ_ε from Example 3.

(i) Example 2 shows that $\tau_\varepsilon < \infty$ (a.s.). Prove that $\tau_\varepsilon \to \infty$ (a.s.) as $\varepsilon \downarrow 0$, so that the probability that the two-dimensional Wiener process hits a particular point is zero even though it hits any neigborhood of this point with probability one.

(ii) Use the method in Example 3 to show that for $d = 2$ and $0 < \varepsilon < |x_0|$

$$\ln|w_{t\wedge\tau_\varepsilon} - x_0| = \ln|x_0| + \int_0^t I_{s\le\tau_\varepsilon}|w_s - x_0|^{-2}(w_s - x_0)\,dw_s.$$

Let $\varepsilon \downarrow 0$ here, and by using (i) conclude that $|w_t - x_0|^{-2}(w_t - x_0) \in \mathcal{S}$ and

$$\ln|w_t - x_0| = \ln|x_0| + \int_0^t |w_s - x_0|^{-2}(w_s - x_0)\,dw_s. \qquad (1)$$

(iii) Prove that $E\ln|w_t - x_0| > \ln|x_0|$ for $t > 0$, so that the stochastic integral in (1) is not a martingale.

8. Girsanov's theorem

Itô's formula allows one to obtain an extremely important theorem about change of probability measure. We consider here a d-dimensional Wiener process $(w_t, \mathcal{F}_t)$ given on a complete probability space $(\Omega, \mathcal{F}, P)$ and assume that the $\mathcal{F}_t$ are complete.

We need the following lemma in which, in particular, we show how one can do Exercises 3.2.5 and 4.3 by using Itô's formula.

1. Lemma. *Let $b \in \mathcal{S}$ be an $\mathbb{R}^d$-valued process. Denote*

$$\rho_t = \rho_t(b) = \exp\Big(\int_0^t b_s\,dw_s - (1/2)\int_0^t (b_s)^2\,ds\Big)$$

$$= \exp\Big(\sum_{i=1}^d \int_0^t b_s^i\,dw_s^i - (1/2)\sum_{i=1}^d \int_0^t (b_s^i)^2\,ds\Big). \tag{1}$$

Then

(i) $d\rho_t = b_t\rho_t\,dw_t$;

(ii) ρ_t *is a supermartingale;*

(iii) *if the process b_t is bounded, then ρ_t is a martingale and, in particular, $E\rho_t = 1$;*

(iv) *if $T \in [0,\infty)$ and $E\rho_T = 1$, then $(\rho_t, \mathcal{F}_t)$ is a martingale for $t \in [0,T]$, and also for any sequence of* bounded $b^n \in \mathcal{S}$ *such that $\int_0^T |b_s^n - b_s|^2\,ds \to 0$ (a.s.) we have*

$$E|\rho_T(b^n) - \rho_T(b)| \to 0. \tag{2}$$

Proof. Assertion (i) follows at once from Itô's formula. To prove (ii) define

$$\tau_n = \inf\{t \geq 0 : \int_0^t |b_s|^2\rho_s^2\,ds \geq n\}.$$

Then $I_{t<\tau_n}b_t\rho_t \in H$ (see the beginning of Sec. 3), and so $\int_0^t I_{s<\tau_n}b_s\rho_s\,dw_s$ is a martingale. By adding that

$$\rho_{t\wedge\tau_n} = 1 + \int_0^{t\wedge\tau_n} b_s\rho_s\,dw_s = 1 + \int_0^t I_{s<\tau_n}b_s\rho_s\,dw_s,$$

we see that $\rho_{t\wedge\tau_n}$ is a martingale. Consequently, for $t_1 \geq t_2$ (a.s.)

$$E(\rho_{t_2 \wedge \tau_n} | \mathcal{F}_{t_1}) = \rho_{t_1 \wedge \tau_n}.$$

As $n \to \infty$, we have $\tau_n \to \infty$ and $t_i \wedge \tau_n \to t_i$, so that by Fatou's theorem (a.s.)

$$E(\rho_{t_2} | \mathcal{F}_{t_1}) \le \rho_{t_1}.$$

This proves (ii) and implies that

$$E \exp\Big(\int_0^t b_s \, dw_s - (1/2) \int_0^t |b_s|^2 \, ds \Big) \le 1. \tag{3}$$

To prove (iii) let $|b_s| \le K$, where K is a constant, and notice that by virtue of (3)

$$E \int_0^t |b_s|^2 \rho_s^2 \, ds \le K^2 E \int_0^t \rho_s^2 \, ds$$

$$= K^2 \int_0^t E\rho_s^2(2b) \exp\Big(\int_0^t |b_s|^2 \, ds \Big) \, ds \le K^2 \int_0^t e^{K^2 s} \, ds < \infty.$$

Hence

$$\int_0^t b_s \rho_s \, dw_s \quad \text{and} \quad \rho_t = 1 + \int_0^t b_s \rho_s \, dw_s$$

are martingales.

To prove (iv), first notice that $E\rho_T(b^n) = 1$ by (iii), $E\rho_T(b) = 1$ by the assumption, and $\rho_T(b^n) \to \rho_T(b)$ in probability by properties of stochastic integrals. This implies (2) by Scheffé's theorem. Furthermore, for $t \le T$ (a.s.)

$$\rho_t(b^n) = E(\rho_T(b^n) | \mathcal{F}_t).$$

Letting $n \to \infty$ here and using Corollary 3.1.10 lead to a similar equality for b in place of b^n, and the martingale property of $\rho_t(b)$ for $t \le T$ now follows from Exercise 3.2.2. The lemma is proved.

2. Remark. Notice again that ρ_t is a solution of $d\rho_t = b_t \rho_t \, dw_t$. We know that in the usual calculus solutions of $df_t = \alpha f_t \, dt$ (that is, exponential functions) play a very big role. As big a role in stochastic calculus is played by *exponential martingales* $\rho_t(b)$.

Inequality (3) implies the following.

3. Corollary. *If b_s is a bounded process or $\int_0^t |b_s|^2 \, ds$ is bounded, then*

$$E \exp \int_0^t b_s \, dw_s < \infty.$$

4. Exercise. (i) By following the argument in the proof of Lemma 1 (ii), prove that if $E\sup_{t\le T}\rho_t<\infty$, then $(\rho_t,\mathcal{F}_t)$ is a martingale for $t\in[0,T]$.

(ii) Use the result of (i) to prove that, if $p>1$ and $N<\infty$ and $E\rho^p_\tau\le N$ for every stopping time $\tau\le T$, then $(\rho_t,\mathcal{F}_t)$ is a martingale for $t\in[0,T]$.

5. Exercise. Use Hölder's inequality and Exercise 4 (ii) to prove that if

$$E\exp\int_0^T c|b_t|^2\,dt<\infty$$

for a constant $c>1/2$, then $E\rho_T(b)=1$.

6. Exercise. By using Exercise 5 and inspecting the inequality

$$1=E\rho_T((1-\varepsilon)b)\le\big[E\rho_T(b)\big]^{1-\varepsilon}\big[E\exp\frac{1-\varepsilon}{2}\int_0^T|b_t|^2\,dt\big]^{\varepsilon},$$

improve the result of Exercise 5 and show that it holds if

$$\lim_{\varepsilon\downarrow 0}\varepsilon\ln E\exp\frac{1-\varepsilon}{2}\int_0^T|b_t|^2\,dt=0,\tag{4}$$

which is true if, for instance, $E\exp(1/2)\int_0^T|b_t|^2\,dt<\infty$ (A. Novikov). It turns out that condition (4) can be relaxed even further by replacing $=0$ with $<\infty$ on the right and lim with $\underline{\lim}$ on the left.

The next lemma treats $\rho_t(b)$ for complex-valued d-dimensional b_t. In this situation we introduce $\rho_t(b)$ by the same formula (1) and for d-vectors $f=(f_1,...,f_d)$ with complex entries f_k denote $(f)^2=\sum_k f_k^2$.

7. Lemma. *If b_t is a bounded d-dimensional* complex-valued *process of class $\mathcal{S}$, then $\rho_t(b)$ is a (complex-valued) martingale and, in particular, $E\rho_t(b)=1$ for any t.*

Proof. Take $t_2>t_1\ge 0$ and $A\in\mathcal{F}_{t_1}$. To prove the lemma it suffices to prove that, if f_t and g_t are bounded $\mathbb{R}^d$-valued processes of class $\mathcal{S}$, then for all complex z

$$EI_A\exp\Big(\int_0^{t_2}(f_s+zg_s)\,dw_s-(1/2)\int_0^{t_2}(f_s+zg_s)^2\,ds\Big)$$

$$=EI_A\exp\Big(\int_0^{t_1}(f_s+zg_s)\,dw_s-(1/2)\int_0^{t_1}(f_s+zg_s)^2\,ds\Big).\tag{5}$$

Observe that (5) holds for real z by Lemma 1 (iii). Therefore we will prove (5) if we prove that both sides are analytic functions of z. In turn to prove this it suffices to show that both sides are continuous and their integrals along closed bounded paths vanish. Finally, due to the analyticity of the expressions under expectation signs and Fubini's theorem we only need to show that, for every $R \in [0, \infty)$ and all $|z| \le R$, these expressions are bounded by a summable function independent of z. This boundedness follows easily from Corollary 3, boundedness of f, g, and the fact that

$$\Big|\exp \int_0^{t_j} (f_s + z g_s)\, dw_s\Big| = \exp \int_0^{t_j} (f_s + g_s \mathrm{Re}\, z)\, dw_s$$

$$\le \exp \int_0^{t_j} (f_s + R g_s)\, dw_s + \exp \int_0^{t_j} (f_s - R g_s)\, dw_s,$$

where we have used the inequality

$$e^{\alpha} \le e^{\alpha} + e^{-\alpha} \le e^{\beta} + e^{-\beta}$$

if $|\alpha| \le |\beta|$. The lemma is proved.

8. Theorem (Girsanov). *Let $T \in [0, \infty)$, and let b be an $\mathbb{R}^d$-valued process of class $\mathcal{S}$ satisfying*

$$E\rho_T(b) = 1.$$

On the measurable space $(\Omega, \mathcal{F})$ introduce the measure $\tilde{P}$ by

$$\tilde{P}(d\omega) = \rho_T(b)(\omega)\, P(d\omega).$$

Then $(\Omega, \mathcal{F}, \tilde{P})$ is a probability space and w_t is a d-dimensional Wiener process on $(\Omega, \mathcal{F}, \tilde{P})$ for $t \le T$.

Proof. That $(\Omega, \mathcal{F}, \tilde{P})$ is a probability space follows from

$$\tilde{P}(\Omega) = \int_{\Omega} \rho_T(b)\, P(d\omega) = E\rho_T(b) = 1.$$

Next denote $\xi_t = w_t - \int_0^t b_s\, ds$. Since $\xi_0 = 0$ and ξ_t is continuous in t, to prove that ξ_t is a Wiener process, it suffices to show that relative to $(\Omega, \mathcal{F}, \tilde{P})$ the joint distributions of the increments of the ξ_t, $t \le T$, are the same as for w_t relative to $(\Omega, \mathcal{F}, P)$.

Let $0 \le t_0 \le t_1 \le ... \le t_n = T$. Fix $\lambda_j \in \mathbb{R}^d$, $j = 0, ..., n-1$, and define the function λ_s as $i\lambda_j$ on $[t_j, t_{j+1})$, $j = 0, ..., n-1$. Also denote by $\tilde{E}$ the expectation sign relative to $\tilde{P}$. By Lemma 7, if b is bounded, then

$$\tilde{E}\exp i\sum_{j=0}^{n-1}\lambda_j(\xi_{t_{j+1}}-\xi_{t_j})=E\exp\Big(\int_0^T\lambda_s\,dw_s-\int_0^T\lambda_s\cdot b_s\,ds\Big)\rho_T(b)$$

$$=E\rho_T(\lambda+b)e^{(1/2)\int_0^T(\lambda_s)^2\,ds}=e^{(1/2)\int_0^T(\lambda_s)^2\,ds}.$$

It follows that

$$\tilde{E}\exp i\sum_{j=0}^{n-1}\lambda_j(\xi_{t_{j+1}}-\xi_{t_j})=\exp\Big(-(1/2)\sum_{j=0}^{n-1}|\lambda_j|^2(t_{j+1}-t_j)\Big).\qquad(6)$$

This proves the theorem if b is bounded. In the general case take a sequence of bounded $b^n\in\mathcal{S}$ such that (a.s.) $\int_0^T|b^n_s-b_s|^2\,ds\to 0$ (for instance, cutting off large values of $|b_s|$). Then

$$E\rho_T(\lambda+b)=\lim_{n\to\infty}E\rho_T(\lambda+b^n),$$

since by Lemma 1 (iv) and the dominated convergence theorem (remember λ_s is imaginary)

$$E|\rho_T(\lambda+b^n)-\rho_T(\lambda+b)|$$

$$=e^{(1/2)\int_0^T|\lambda_s|^2\,ds}E\big|\rho_T(b^n)e^{-\int_0^T\lambda_s\cdot b^n_s\,ds}-\rho_T(b)e^{-\int_0^T\lambda_s\cdot b_s\,ds}\big|$$

$$\le e^{(1/2)\int_0^T|\lambda_s|^2\,ds}\big(E|\rho_T(b^n)-\rho_T(b)|$$

$$+E\big|e^{-\int_0^T\lambda_s\cdot b_s\,ds}-e^{-\int_0^T\lambda_s\cdot b^n_s\,ds}\big|\rho_T(b)\big)\to 0.$$

This and (6) yield the result in the general case. The theorem is proved.

Girsanov's theorem and the lemmas proved before it have numerous applications. We discuss only few of them.

From the theory of ODE's it is known that the equation $dx_t=b(t,x_t)\,dt$ need not have a solution for any bounded Borel b. In contrast with this it turns out that, for almost any trajectory of the Wiener process, the equation $dx_t=b(t,x_t+w_t)\,dt$ does have a solution whenever b is Borel and bounded. This fact is obtained from the following theorem after replacing x_t with ξ_t-w_t.

9. Theorem. *Let $b(t,x)$ be an $\mathbb{R}^d$-valued Borel bounded function on $(0,\infty)\times\mathbb{R}^d$. Then there exist a probability space $(\Omega,\mathcal{F},P)$, a d-dimensional continuous process ξ_t and a d-dimensional Wiener process w_t defined on that space for $t\in[0,T]$ such that*

$$\xi_t=\int_0^t b(s,\xi_s)\,ds+w_t \tag{7}$$

for all $t\in[0,T]$ and $\omega\in\Omega$.

Proof. Take any complete probability space $(\Omega,\mathcal{F},\tilde{P})$ carrying a d-dimensional Wiener process, say ξ_t. Define

$$w_t=\xi_t-\int_0^t b(s,\xi_s)\,ds$$

and on $(\Omega,\mathcal{F})$ introduce a new measure P by the formula

$$P(d\omega)=\exp\Big(\int_0^T b(s,\xi_s)\,d\xi_s-(1/2)\int_0^T |b(s,\xi_s)|^2\,ds\Big)\tilde{P}(d\omega).$$

Then $(\Omega,\mathcal{F},P)$ is a probability space, w_t is a Wiener process on $(\Omega,\mathcal{F},P)$ for $t\in[0,T]$, and, by definition, ξ_t solves (7). The theorem is proved.

The proof of this theorem looks like a trick and usually leaves the reader unsatisfied. Indeed firstly, no real method is given such as Picard's method of successive approximations or Euler's method allowing one to find solutions. Secondly, the question remains as to whether one can find solutions on a *given* probability space without changing it, so that ξ_t would be defined by the Wiener process w_t and not conversely. Theorem 9 was proved by I. Girsanov around 1965. Only in 1978 did A. Veretennikov prove that indeed the solutions can be found on any probability space, and only in 1996 did it become clear that Euler's method allows one to construct the solutions effectively.

Let us also show the application of Girsanov's theorem to finding

$$P(\max_{t\le 1}(w_t+t)\ge 1),$$

where w_t is a one-dimensional Wiener process. Let $b=-1$ and

$$\tilde{P}(d\omega)=e^{-w_t-1/2}P(d\omega).$$

By Girsanov's theorem $\bar{w}_t:=w_t+t$ is a Wiener process for $t\in[0,1]$. Since the distributions of Wiener processes in the space of continuous functions are all the same and are given by Wiener measure, we conclude

$$P(\max_{t\leq 1}(w_t+t)\geq 1)=\int_\Omega I_{\max_{t\leq 1}\bar{w}_t\geq 1}e^{\bar{w}_1-1/2}e^{-w_1-1/2}P(d\omega)$$

$$=\int_\Omega I_{\max_{t\leq 1}\bar{w}_t\geq 1}e^{\bar{w}_1-1/2}\tilde{P}(d\omega)=EI_{\max_{t\leq 1}w_t\geq 1}e^{w_1-1/2}.$$

Now remember the result of Exercise 2.2.10, which is

$$P(\max_{t\leq 1}w_t\geq 1, w_1\leq x)=\begin{cases}P(w_1\geq 2-x) & \text{if} \quad x\leq 1,\\ 2P(w_1\geq 1)-P(w_1\geq x) & \text{if} \quad x\geq 1.\end{cases}$$

Then by using the hint to Exercise 2.2.12, we get

$$P(\max_{t\leq 1}(w_t+t)\geq 1)=\int_1^\infty e^{x-1/2}\frac{1}{\sqrt{2\pi}}e^{-x^2/2}\,dx$$

$$+\int_{-\infty}^1 e^{x-1/2}\frac{1}{\sqrt{2\pi}}e^{-(2-x)^2/2}\,dx=\frac{1}{\sqrt{2\pi e}}\int_1^\infty(e^x+e^{2-x})e^{-x^2/2}\,dx.$$

In the following exercise we suggest the reader derive a particular case of the Burkholder-Davis-Gundy inequalities.

10. Exercise. Let τ be a bounded stopping time. Then for any real λ we have

$$Ee^{\lambda w_\tau-\lambda^2\tau/2}=1.$$

By using Corollary 3, prove that we can differentiate this equality with respect to λ as many times as we wish, bringing all derivatives inside the expectation sign. Then, for any integer $k\geq 1$, prove that

$$E(a_0w_\tau^{2k}+a_2w_\tau^{2k-2}\tau+a_4w_\tau^{2k-4}\tau^2+\ldots+a_{2k}\tau^k)=0,$$

where $a_0,\ldots,a_{2k}$ are certain absolute constants (depending on k) and $a_0\neq 0$ and $a_{2k}\neq 0$. Finally, remembering Hölder's inequality, prove that

$$Ew_\tau^{2k}\leq NE\tau^k,\quad E\tau^k\leq NEw_\tau^{2k},$$

where the constant N depends only on k.

9. Stochastic Itô equations

A very wide class of continuous stochastic processes can be obtained by modeling various diffusion processes. They are generally characterized by being Markov and having local drift and diffusion; that is, behaving near a point x on the time interval Δt like $\sigma(x)\,\Delta w_t + b(x)\,\Delta t$, where $\sigma(x)$ is the local diffusion coefficient and $b(x)$ is the local drift. A quite satisfactory model of such processes is given by solutions of stochastic Itô equations.

Let $(\Omega, \mathcal{F}, P)$ be a complete probability space, $(w_t, \mathcal{F}_t)$ a d-dimensional Wiener process given for $t \geq 0$. Assume that the σ-fields $\mathcal{F}_t$ are complete (which is needed, for instance, to define stochastic integrals as continuous $\mathcal{F}_t$-adapted processes). Let $b(t,x)$ and $\sigma(t,x)$ be Borel functions defined on $(0,\infty) \times \mathbb{R}^d$. We assume that b is $\mathbb{R}^{d_1}$-valued and σ takes values in the set of $d_1 \times d$ matrices. Finally, assume that there exists a constant $K < \infty$ such that for all x, y, t

$$\begin{aligned} ||\sigma(t,x)|| + |b(t,x)| &\leq K(1+|x|), \\ ||\sigma(t,x) - \sigma(t,y)|| + |b(t,x) - b(t,y)| &\leq K|x-y|, \end{aligned} \tag{1}$$

where by $||\sigma||$ for a matrix σ we mean $\big(\sum_{i,j}(\sigma^{ij})^2\big)^{1/2}$.

Take an $\mathcal{F}_0$-measurable $\mathbb{R}^{d_1}$-valued random variable ξ_0 and consider the following Itô equation:

$$\xi_t = \xi_0 + \int_0^t \sigma(s,\xi_s)\,dw_s + \int_0^t b(s,\xi_s)\,ds \quad t \geq 0. \tag{2}$$

By a solution of this equation we mean a continuous $\mathcal{F}_t$-adapted process given for $t \geq 0$ and such that (2) holds for all $t \geq 0$ at once with probability one. Notice that, for any continuous $\mathcal{F}_t$-adapted process ξ_t given for $t \geq 0$, the function $\xi(\omega,t)$ is jointly measurable in (ω,t) and the functions $\sigma(t,\xi_t)$ and $b(t,\xi_t)$ are jointly measurable in (ω,t) and $\mathcal{F}_t$-adapted. In addition, $\sigma(t,\xi_t)$ and $b(t,\xi_t)$ are bounded for each ω on $[0,t]$ for any $t < \infty$. It follows that for such processes ξ_t the right-hand side of (2) makes sense.

In our investigation of solvability of (2) we use the following lemma, in which $\mathfrak{M}$ is *the set of all finite stopping times*.

1. Lemma. (i) *Let ξ_t and η_t be continuous nonnegative $\mathcal{F}_t$-adapted processes, $f \in \mathcal{S}$, and*

$$\eta_t \leq \xi_t + \int_0^t f_s\,dw_s.$$

Let ξ_t be nondecreasing in t and $E\xi_\tau < \infty$ for every $\tau \in \mathfrak{M}$. Then

$$E\eta_\tau \le E\xi_\tau \quad \forall \tau \in \mathfrak{M}.$$

(ii) *Let η_t be a continuous nonnegative $\mathcal{F}_t$-adapted process and $E\eta_\tau \le N$ for all $\tau \in \mathfrak{M}$, where N is a constant (independent of τ). Then, for every $\varepsilon > 0$,*

$$P(\sup_t \eta_t > \varepsilon) \le N/\varepsilon.$$

Proof. (i) Denote

$$\tau_n = \inf\{t \ge 0 : \int_0^t |f_s|^2 \, ds \ge n\}.$$

Then $\tau_n \uparrow \infty$ as $n \to \infty$ and $I_{s<\tau_n} f_s \in H$. Hence, for every $\tau \in \mathfrak{M}$,

$$\eta_{\tau \wedge \tau_n} \le \xi_\tau + \int_0^\tau I_{s<\tau_n} f_s \, dw_s, \quad E\eta_{\tau\wedge\tau_n} \le E\xi_\tau.$$

After that, Fatou's theorem proves (i).

(ii) Define

$$\tau = \inf\{t \ge 0 : \eta_t \ge \varepsilon\}.$$

Then

$$P(\sup_t \eta_t > \varepsilon) \le P(\tau < \infty) = \lim_{t\to\infty} P(\tau < t)$$

$$\le \varliminf_{t\to\infty} P(\eta_{t\wedge\tau} \ge \varepsilon) \le \varliminf_{t\to\infty} \frac{1}{\varepsilon} E\eta_{t\wedge\tau} \le \frac{N}{\varepsilon}.$$

The lemma is proved.

2. Theorem. *Equation (2) has a solution.*

Proof. We apply the usual Picard method of successive approximations. For $n \ge 0$ define

$$\xi_t(n+1) = \xi_0 + \int_0^t \sigma(s, \xi_s(n)) \, dw_s + \int_0^t b(s, \xi_s(n)) \, ds, \quad \xi_t(0) \equiv \xi_0. \quad (3)$$

Notice that all the processes $\xi_t(n)$ are continuous and $\mathcal{F}_t$-adapted, and our definition makes sense for all $n \ge 0$. Define

$$\psi_t = e^{-N_0 t - |\xi_0|},$$

where the constant $N_0 \geq 1$ will be specified later. We want to show by induction that

$$\sup_{\tau \in \mathfrak{M}} E\big(\psi_\tau |\xi_\tau(n)|^2 + \int_0^\tau \psi_s |\xi_\tau(n)|^2 \, ds\big) < \infty. \tag{4}$$

For $n = 0$ estimate (4) is obvious, since $a^2 e^{-|a|}$ is a bounded function and $\int_0^\infty e^{-N_0 t} \, dt < \infty$. Next, by Itô's formula we find that

$$d(\psi_t |\xi_t(n+1)|^2) = |\xi_t(n+1)|^2 d\psi_t + \psi_t d|\xi_t(n+1)|^2$$

$$= \psi_t[-N_0 |\xi_t(n+1)|^2 + 2\xi_t(n+1) \cdot b(t, \xi_t(n)) + ||\sigma(t, \xi_t(n))||^2] \, dt$$

$$+ 2\psi_t \sigma^*(t, \xi_t(n)) \xi_t(n+1) \, dw_t.$$

Here to estimate the expression in the brackets we use $2ab \leq a^2 + b^2$ and (1) to find that

$$2\xi_t(n+1) \cdot b(t, \xi_t(n)) + ||\sigma(t, \xi_t(n))||^2$$

$$\leq |\xi_t(n+1)|^2 + |b(t, \xi_t(n))|^2 + ||\sigma(t, \xi_t(n))||^2$$

$$\leq |\xi_t(n+1)|^2 + 2K^2(1 + |\xi_t(n)|)^2 \leq |\xi_t(n+1)|^2 + 4K^2(1 + |\xi_t(n)|^2).$$

Hence, for $N_0 \geq 2$

$$\psi_t |\xi_t(n+1)|^2 + \int_0^t \psi_s |\xi_s(n+1)|^2 \, ds \leq \psi_0 |\xi_0|^2 + 4K^2 \int_0^t (1 + |\xi_s(n)|^2) \psi_s \, ds$$

$$+ 2 \int_0^t \psi_s \sigma^*(s, \xi_s(n)) \xi_s(n+1) \, dw_s$$

Applying Lemma 1 leads to (4).

Further,

$$d(\psi_t |\xi_t(n+1) - \xi_t(n)|^2) = \psi_t[-N_0 |\xi_t(n+1) - \xi_t(n)|^2$$

$$+ 2(\xi_t(n+1) - \xi_t(n)) \cdot (b(t, \xi_t(n)) - b(t, \xi_t(n-1)))$$

$$+ ||\sigma(t, \xi_t(n)) - \sigma(t, \xi_t(n-1))||^2] \, dt$$

$$+ 2\psi_t [\sigma(t, \xi_t(n)) - \sigma(t, \xi_t(n-1))]^* (\xi_t(n+1) - \xi_t(n)) \, dw_t.$$

Due to (1) the expression in the brackets is less than

$$-(N_0 - 1)|\xi_t(n+1) - \xi_t(n)|^2 + 2K^2 |\xi_t(n) - \xi_t(n-1)|^2.$$

Now we make the final choice of N_0 and take it equal to $4K^2 + 2$, so that $N_0 \geq 2$ as we needed above and $c := N_0 - 1 \geq c/2 \geq 2K^2$. Then we get

$$d(\psi_t|\xi_t(n+1)-\xi_t(n)|^2)+c\psi_t|\xi_t(n+1)-\xi_t(n)|^2\,dt$$
$$\leq (c/2)\psi_t|\xi_t(n)-\xi_t(n-1)|^2\,dt$$
$$+2\psi_t[\sigma(t,\xi_t(n))-\sigma(t,\xi_t(n-1))]^*(\xi_t(n+1)-\xi_t(n))\,dw_t.$$

It follows by Lemma 1 that for any $\tau \in \mathfrak{M}$

$$E\psi_\tau|\xi_\tau(n+1)-\xi_\tau(n)|^2+cE\int_0^\tau \psi_t|\xi_t(n+1)-\xi_t(n)|^2\,dt$$

$$\leq (c/2)E\int_0^\tau \psi_t|\xi_t(n)-\xi_t(n-1)|^2\,dt. \tag{5}$$

By iterating (5) we get

$$E\int_0^\tau \psi_t|\xi_t(n+1)-\xi_t(n)|^2\,dt \leq 2^{-n}E\int_0^\infty \psi_t|\xi_t(1)-\xi_t(0)|^2\,dt =: N2^{-n}.$$

Coming back to (5), we now see that

$$E\psi_\tau|\xi_\tau(n+1)-\xi_\tau(n)|^2 \leq cN2^{-n},$$

which by Lemma 1 yields

$$P\{\sup_{t\geq 0}(\psi_t|\xi_t(n+1)-\xi_t(n)|^2)\geq n^{-4}\}\leq n^4cN2^{-n}.$$

By the Borel-Cantelli lemma we conclude that the series

$$\sum_{n=1}^\infty \psi_t^{1/2}|\xi_t(n+1)-\xi_t(n)|$$

converges uniformly on $[0,\infty)$ with probability one. Obviously this implies that $\xi_t(n)$ converges uniformly on each finite time interval with probability one. Let ξ_t denote the limit. Then, by the dominated convergence theorem (or just because of the uniform convergence to zero of the integrands),

$$\int_0^t ||\sigma(s,\xi_s(n))-\sigma(s,\xi_s)||^2\,ds+\int_0^t |b(s,\xi_s(n))-b(s,\xi_s)|\,ds \to 0$$

(a.s.). Furthermore, ξ_t is continuous in t and $\mathcal{F}_t$-adapted. Therefore, by letting $n\to\infty$ in (3) we obtain (a.s.)

$$\xi_t=\xi_0+\int_0^t \sigma(s,\xi_s)\,dw_s+\int_0^t b(s,\xi_s)\,ds.$$

The theorem is proved.

It is convenient to deduce the uniqueness of solutions to (2) from the following theorem on continuous dependence of solutions on initial data.

3. Theorem. *Let $\mathcal{F}_0$-measurable d-vector valued random variables ξ_0^n satisfy $\xi_0^n \to \xi_0$ (a.s.) as $n \to \infty$. Let (a.s.) for $t \geq 0$*

$$\xi_t = \xi_0 + \int_0^t \sigma(s, \xi_s)\, dw_s + \int_0^t b(s, \xi_s)\, ds,$$

$$\xi_t^n = \xi_0^n + \int_0^t \sigma(s, \xi_s^n)\, dw_s + \int_0^t b(s, \xi_s^n)\, ds.$$

Then

$$P\{\sup_{t \leq T} |\xi_t^n - \xi_t| \geq \varepsilon\} \to 0$$

as $n \to \infty$ for any $\varepsilon > 0$ and $T \in [0, \infty)$.

Proof. Take N_0 from the previous proof and denote

$$\psi_t = \exp(-N_0 t - \sup_n |\xi_0^n|).$$

Notice that $|\xi_0| \leq \sup_n |\xi_0^n|$ (a.s.). Also the last sup is finite, and hence $\psi_t > 0$ (a.s.). By Itô's formula

$$d(\psi_t |\xi_t - \xi_t^n|^2) = \psi_t[-N_0 |\xi_t - \xi_t^n|^2 + 2(\xi_t - \xi_t^n) \cdot (b(t, \xi_t) - b(t, \xi_t^n))$$

$$+ ||\sigma(t, \xi_t) - \sigma(t, \xi_t^n)||^2]\, dt + 2[\sigma(t, \xi_t) - \sigma(t, \xi_t^n)]^* (\xi_t - \xi_t^n)\, dw_t.$$

By following the proof of Theorem 2 we see that the expression in brackets is nonpositive. Hence for any $\tau \in \mathfrak{M}$

$$E\psi_\tau |\xi_\tau - \xi_\tau^n|^2 \leq E\psi_0 |\xi_0 - \xi_0^n|^2.$$

Here the random variables $\psi_0 |\xi_0 - \xi_0^n|^2$ are bounded by a constant independent of n and tend to zero (a.s.). By the dominated convergence theorem and Lemma 1,

$$P\{\sup_{t \geq 0} \psi_t |\xi_t - \xi_t^n|^2 > \varepsilon\} \leq \frac{1}{\varepsilon} E\psi_0 |\xi_0 - \xi_0^n|^2 \to 0.$$

Consequently, $\sup_{t \geq 0} \psi_t |\xi_t - \xi_t^n|^2$ converges to zero in probability and, since

$$\sup_{t \leq T} |\xi_t - \xi_t^n|^2 \leq \psi_T^{-1} \sup_{t \geq 0} \psi_t |\xi_t - \xi_t^n|^2,$$

the random variables $\sup_{t\leq T}|\xi_t-\xi^n_t|^2$ converge to zero in probability as well. The theorem is proved.

4. Corollary (uniqueness). *If ξ_t and η_t are two solutions of (2), then*

$$P(\sup_{t\geq 0}|\xi_t-\eta_t|>0)=0.$$

The following corollary states the so-called *Feller property* of solutions of (2).

5. Corollary. *For $x\in\mathbb{R}^d$, let $\xi_t(x)$ be a solution of equation (2) with $\xi_0\equiv x$. Then, for every bounded continuous function f and $t\geq 0$, the function $Ef(\xi_t(x))$ is a continuous function of x.*

6. Corollary. *In notation from the proof of Theorem 3,*

$$P\{\psi_T\sup_{t\leq T}|\xi_t-\xi^n_t|^2>\varepsilon\}\leq\frac{1}{\varepsilon}E\psi_0|\xi_0-\xi^n_0|^2.$$

10. An example of a stochastic equation

In one-dimensional space we consider the following equation:

$$\xi_t=\int_0^t\sigma(\xi_s)\,dw_s+\int_0^t b(\xi_s)\,ds \tag{1}$$

with a one-dimensional Wiener process w_t which is Wiener relative to a filtration of complete σ-fields $\mathcal{F}_t$. We assume that σ and b are bounded functions satisfying a Lipschitz condition, so that there exists a unique solution of (1).

Fix $r\leq 0$ and let

$$\tau(r)=\inf\{t\geq 0:\xi_t\notin(r,1)\}.$$

By Exercise 2.4 we have that $\tau(r)$ is a stopping time relative to $\mathcal{F}_t$. We want to find $E\tau(r)$. By Itô's formula, for twice continuously differentiable functions u we have

$$u(\xi_t)=u(0)+\int_0^t Lu(\xi_s)\,ds+\int_0^t\sigma(\xi_s)u'(\xi_s)\,dw_s, \tag{2}$$

where the operator L, called *the generator* of the process ξ_t, is given by

$$Lu(x)=a(x)u''(x)+b(x)u'(x),\quad a=(1/2)\sigma^2.$$

If we can substitute $\tau(r)$ in place of t in (2) and take the expectation of both sides and be sure that the expectation of the stochastic integral vanishes, then we find that

$$u(0) = Eu(\xi_{\tau(r)}) - E\int_0^{\tau(r)} Lu(\xi_s)\,ds. \tag{3}$$

Upon noticing after this that

$$E\tau(r) = E\int_0^{\tau(r)} dt,$$

we arrive at the following way to find $E\tau(r)$: Solve the equation $Lu = -1$ on $(a, 1)$ with boundary conditions $u(r) = u(1) = 0$ (in order to have $u(\xi_{\tau(r)}) = 0$); then $E\tau(r)$ should be equal to $u(0)$.

1. Lemma. *Let $a(x) \geq \delta$, where δ is a constant, $\delta > 0$. For $x, y \in [r, 1]$ define*

$$\phi(x) = \exp\big(-\int_0^x b(s)/a(s)\,ds\big), \quad \psi = \int_r^1 \phi(x)\,dx,$$

$$g(x, y) = \frac{1}{\psi a(y)\phi(y)}\int_{x\vee y}^1 \phi(s)\,ds \int_r^{x\wedge y} \phi(s)\,ds.$$

Then, for any continuous function $f(x)$ given on $[r, 1]$, the function

$$u(x) := \int_r^1 g(x, y)f(y)\,dy \tag{4}$$

is twice continuously differentiable on $[r, 1]$, vanishes at the end points of this interval, and satisfies $Lu = -f$ on $[r, 1]$.

The proof of this lemma is suggested as an exercise, which the reader is supposed to do by solving the equation $au'' + bu' = -f$ on $[r, 1]$ with boundary condition $u(r) = u(1) = 0$ and then transforming the result to the right-hand side of (4).

2. Theorem. *Under the assumptions of Lemma 1, for any Borel nonnegative function f we have*

$$E\int_0^{\tau(r)} f(\xi_t)\,dt = \int_r^1 g(0, y)f(y)\,dy. \tag{5}$$

In particular,

$$E\tau(r) = \int_r^1 g(0,y)\,dy.$$

Proof. A standard measure-theoretic argument shows that it suffices to prove the theorem for nonnegative bounded continuous f. For such a function, define $u(x)$ in $[r,1]$ as a solution to the equation $au'' + bu' = -f$ on $[r,1]$ with boundary condition $u(r) = u(1) = 0$. Then continue u outside $[r,1]$ to get a twice continuously differentiable function on $\mathbb{R}$, and keeping the same notation for the continuation use (2) with $t \wedge \tau(r)$ in place of t. After that take expectations, and notice that the expectation of the stochastic integral vanishes since $u'(\xi_s)$ is bounded on $[0,\tau(r)]$. Then we get

$$Eu(\xi_{t\wedge\tau(r)}) = u(0) - E\int_0^{t\wedge\tau(r)} f(\xi_s)\,ds. \tag{6}$$

If we take $f \equiv 1$ here, then we see that $E(t \wedge \tau(r))$ is bounded by a constant independent of t. It follows by the monotone convergence theorem that $E\tau(r) < \infty$ and $\tau(r) < \infty$ (a.s.). Hence by letting $t \to \infty$ in (6) and noticing that $u(\xi_{t\wedge\tau(r)}) \to u(\xi_{\tau(r)}) = 0$ (a.s.) due to the boundary conditions, by the dominated convergence theorem and the monotone convergence theorem ($f \geq 0$) we get

$$u(0) = E\int_0^{\tau(r)} f(\xi_s)\,ds,$$

which is (5) owing to Lemma 1. The theorem is proved.

As a consequence of (5), as in Exercise 2.6.6, one finds that the average time the process ξ_t spends in an interval $[c,d] \subset (r,1)$ before exiting from $(r,1)$ is given by

$$\int_c^d g(0,y)\,dy.$$

The remaining part of this section is aimed at exhibiting a rather unexpected effect which happens when $b \equiv 1$ and a is very close to zero for $x < 0$ and very large for $x > 0$. It turns out that in this situation $E\tau(r)$ is close to 1, and the average time spent by the process in a very small neighborhood of zero before exiting from $(r,1)$ is also close to 1. Hence the process spends almost all time near the origin and then immediately "jumps" out of $(r,1)$. The unexpected here is that there is the unit drift pushing the particle to the right, and the diffusion is usually expected to get the particle around this deterministic motion but not to practically *stop* it. Furthermore, it turns out that the process spends almost all time in a small region where *the diffusion is small* and, remember, $b \equiv 1$.

The following exercise makes it natural that if the diffusion on $(-\infty, 0)$ is slow then $E\tau(r)$ is close to 1. Assertion (i) in Exercise 3 looks quite natural because neither diffusion vanishing for $x \leq c$ nor positive drift can bring our moving particle ξ_t starting at zero below $c \leq 0$.

3. Exercise. Assume that $\sigma(x) = 0$ for $x \leq c$, where c is a constant, $c \leq 0$ and $b(x) \geq 0$ for all x.

(i) Prove that $\xi_t \geq c$ for all $t \geq 0$ (a.s.).

(ii) Prove that, if $c > r$ and $b \geq \delta$, where δ is a constant and $\delta > 0$, then

$$E \int_0^{\tau(r)} b(\xi_t)\, dt = 1$$

and, in particular, if $b \equiv 1$, then $E\tau(r) = 1$.

4. Exercise*. Let $b \equiv 1$. Prove that $E\tau(r) \leq 1$.

We will be dealing with σ depending on $r \in (-1, 0)$, which will be sent to zero. Let

$$b \equiv 1, \quad a(x) = r^4 \quad \text{if} \quad x < 0, \quad a(x) = |r|^{-1} \quad \text{if} \quad x > |r|,$$

and let a be linear on $[0, |r|]$ with $a(0) = r^4$ and $a(|r|) = |r|^{-1}$, so that a is a Lipschitz continuous function. Naturally, we take $\sigma = \sqrt{2a}$. Then σ is also Lipschitz continuous, and the corresponding process ξ_t is well defined. Now we can make precise what is stated before Exercise 3.

5. Theorem. *As $r \uparrow 0$ we have*

$$E\tau(r) \to 1, \quad E \int_0^{\tau(r)} I_{[r,0]}(\xi_t)\, dt \to 1.$$

Proof. Due to Exercise 4 the first relation follows from the second one, which in turn by virtue of Theorem 2 can be rewritten as

$$\int_r^0 g(0, y)\, dy \to 1. \tag{7}$$

Next, in the notation from Lemma 1 the integral in (7) equals

$$\psi^{-1} \int_0^1 \phi(s)\, ds \int_r^0 \frac{1}{a(y)\phi(y)} \Big(\int_r^y \phi(s)\, ds \Big)\, dy$$

$$= \psi^{-1} \int_0^1 \phi(s)\, ds \int_r^0 \Big(\int_r^y \phi(s)\, ds \Big)\, d\phi^{-1}(y)$$

$$= \psi^{-1} \int_0^1 \phi(s)\, ds \Big[\int_r^0 \phi(s)\, ds - |r| \Big]. \tag{8}$$

Furthermore, $\phi(s) = e^{|s|/r^4}$ if $s \leq 0$, whence

$$\int_r^0 \phi(s)\, ds = \int_r^0 e^{|s|/r^4}\, ds \to \infty$$

as $r \uparrow 0$. To investigate other terms in (8), observe that, for $s \in [0, |r|]$,

$$\int_0^s a^{-1}(t)\, dt = \int_0^s \frac{r^2}{r^6 + (1 - |r|^5)t}\, dt = \frac{r^2}{1 - |r|^5} \ln(1 + (r^{-6} - |r|^{-1})s)$$

$$\leq \frac{r^2}{1 - |r|^5} \ln(1 + (|r|^{-5} - 1)) = \frac{-5r^2}{1 - |r|^5} \ln |r| \to 0.$$

For $s \geq |r|$

$$\int_0^s a^{-1}(t)\, dt \leq \frac{-5r^2}{1 - |r|^5} \ln |r| + \int_{|r|}^s a^{-1}(t)\, dt = \frac{-5r^2}{1 - |r|^5} \ln |r| + (s - |r|)|r| \to 0.$$

It follows that

$$\int_0^1 \phi(s)\, ds \to 1, \quad \psi = \int_r^0 \phi(s)\, ds + \int_0^1 \phi(s)\, ds \sim \int_r^0 \phi(s)\, ds \to \infty,$$

so that indeed the last expression in (8) tends to 1 as $r \uparrow 0$. The theorem is proved.

11. The Markov property of solutions of stochastic equations

1. Definition. A vector-valued random process ξ_t given for $t \geq 0$ is called *Markov* if for every Borel bounded function $f(x)$, $n = 1, 2, \ldots$, and $t, t_1, \ldots, t_n$ such that $0 \leq t_1 \leq \ldots \leq t_n \leq t$ we have (a.s.)

$$E(f(\xi_t) | \xi_{t_1}, \ldots, \xi_{t_n}) = E(f(\xi_t) | \xi_{t_n}).$$

Remember that $E(f(\xi_t)|\xi_{t_1},...,\xi_{t_n})$ was defined as the conditional expectation of $f(\xi_t)$ given the σ-field $\sigma(\xi_{t_1},...,\xi_{t_n})$ generated by $\xi_{t_1},...,\xi_{t_n}$. We know that $E(f(\xi_t)|\xi_{t_1},...,\xi_{t_n})$ is the best (in the mean square sense) estimate of $f(\xi_t)$ which can be constructed on the basis of $\xi_{t_1},...,\xi_{t_n}$. If we treat this estimate as the prediction of $f(\xi_t)$ on the basis of $\xi_{t_1},...,\xi_{t_n}$, then we see that for Markov processes there is no need to remember the past to predict the future: remembering the past does not affect our best prediction.

In this section we make the same assumptions as in Sec. 9, and first we try to explain why the solution of equation (9.2) should possess the Markov property.

Let $0 \leq t_1 \leq ... \leq t_n$. Obviously, for $t \geq t_n$ the process ξ_t satisfies

$$\xi_t = \xi_{t_n} + \int_{t_n}^t \sigma(s,\xi_s)\,dw_s + \int_{t_n}^t b(s,\xi_s)\,ds.$$

This makes it more or less clear that ξ_t is completely defined by ξ_{t_n} and the increments of $w.$ after time t. For t fixed one may write this fact as

$$\xi_t = g(\xi_{t_n}, w_u - w_v, u \geq v \geq t_n), \quad f(\xi_t) = h(\xi_{t_n}, w_u - w_v, u \geq v \geq t_n).$$

Next, observe that ξ_t is $\mathcal{F}_t$-measurable and $w_u - w_v$ is independent of $\mathcal{F}_t$ by definition. Then we see that $w_u - w_v$ is independent of ξ_t if $u \geq v \geq t$, and Theorem 3.1.13 seems to imply that

$$E(f(\xi_t)|\xi_{t_1},...,\xi_{t_n}) = E(h(\xi_{t_n}, w_u - w_v, u \geq v \geq t_n)|\xi_{t_1},...,\xi_{t_n})$$

$$= Eh(x, w_u - w_v, u \geq v \geq t_n)|_{x=\xi_{t_n}} \tag{1}$$

(a.s.). Since one gets the same result for $E(f(\xi_t)|\xi_{t_n})$, we see that ξ_t is a Markov process. Unfortunately this very convincing explanation cannot count as a proof, since to apply Theorem 3.1.13 in (1) we have to know that $h(x, w_u - w_v, u \geq v \geq t_n)$ is measurable with respect to (x,ω). Actually, on the basis of Kolmogorov's theorem for random fields one can prove that $g(x, w_u - w_v, u \geq v \geq t_n)$ has a modification which is continuous in x, so that $h(x, w_u - w_v, u \geq v \geq t_n)$ has a modification measurable with respect to (x,ω). However we prefer a different way of proving the Markov property, because it is shorter and applicable in many other situations.

Let us fix x and consider the equation

$$\xi_t = x + \int_{t_n}^t \sigma(s,\xi_s)\,dw_s + \int_{t_n}^t b(s,\xi_s)\,ds, \quad t \geq t_n. \tag{2}$$

Above we have investigated such equations only with $t_n = 0$. The case $t_n \geq 0$ is not any different. Therefore, for $t \geq t_n$ equation (2) has a continuous $\mathcal{F}_t$-adapted solution. We denote this solution by $\xi_t(x)$. As in Theorem 9.3, one proves that $\xi_t(x_n) \xrightarrow{P} \xi_t(x)$ if $x_n \to x$. Among other things, this implies the uniqueness of solutions to (2).

2. Lemma. *For any $t \geq t_n$ and $x \in \mathbb{R}^{d_1}$ the random variable $\xi_t(x)$ is measurable with respect to the completion of the σ-field $\mathcal{G}_t$ generated by $w_s - w_{t_n}$, $s \in [t, t_n]$.*

Proof. It is easy to understand that the process $\bar{w}_t := w_{t+t_n} - w_{t_n}$ is a Wiener process and, by definition, $\mathcal{G}_t = \mathcal{F}^{\bar{w}}_{t-t_n}$. Let $\bar{\mathcal{S}}$ be the set $\mathcal{S}$ constructed from $\bar{\mathcal{F}}^{\bar{w}}_t := (\mathcal{F}^{\bar{w}}_t)^P$. It turns out that, for any $\mathbb{R}^{d_1}$-valued process $\zeta \in \bar{\mathcal{S}}$, we have $I_{t \geq t_n}\zeta_{t-t_n} \in \mathcal{S}$ and (a.s.)

$$\int_0^t \zeta_s \, d\bar{w}_s = \int_{t_n}^{t_n+t} \zeta_{s-t_n} \, dw_s \quad \forall t \geq 0. \tag{3}$$

Here the first statement is obvious since $\bar{\mathcal{F}}^{\bar{w}}_t \subset \mathcal{F}_{t+t_n}$ and an $\bar{\mathcal{F}}^{\bar{w}}_t$-measurable variable ζ_t is also $\mathcal{F}_{t+t_n}$-measurable. In (3) both sides are continuous in t. Therefore, it suffices to prove (3) for each particular t. Since the equality is obvious for step functions, a standard argument applied already at least twice in previous sections proves (3) in the general case.

Next, by Theorem 9.2 there is an $\bar{\mathcal{F}}^{\bar{w}}_t$-adapted solution to

$$\bar{\xi}_t = x + \int_0^t \sigma(s + t_n, \bar{\xi}_s) \, d\bar{w}_s + \int_0^t b(s + t_n, \bar{\xi}_s) \, ds.$$

By virtue of (3) the process $\bar{\xi}_{t-t_n}$ satisfies (2) for $t \geq t_n$ and is $\mathcal{F}_t$-adapted. It follows from uniqueness that $\xi_t(x) = \bar{\xi}_{t-t_n}$ (a.s.), and since $\bar{\xi}_{t-t_n}$ is $\bar{\mathcal{F}}^{\bar{w}}_{t-t_n}$-measurable (that is, $\bar{\mathcal{G}}_t$-adapted) the lemma is proved.

3. Lemma. *The σ-fields $\mathcal{G}_t$ and $\mathcal{F}_{t_n}$ are independent. That is, $P(AB) = P(A)P(B)$ for each $A \in \mathcal{G}_t$ and $B \in \mathcal{F}_{t_n}$.*

Proof. Let $B \in \mathcal{F}_{t_n}$, Borel $\Gamma_1, ..., \Gamma_m \subset \mathbb{R}^{d_1}$, and $0 \leq s_1 \leq ... \leq s_m$. By using properties of conditional expectations we find that

$$P\{B, (\bar{w}_{s_1}, \bar{w}_{s_2} - \bar{w}_{s_1}, ..., \bar{w}_{s_m} - \bar{w}_{s_{m-1}}) \in \Gamma_1 \times ... \times \Gamma_m\}$$

$$= EI_B I_{\Gamma_1}(w_{t_n+s_1} - w_{t_n}) \cdot ... \cdot I_{\Gamma_{m-1}}(w_{t_n+s_{m-1}} - w_{t_n+s_{m-2}})$$

$$\times E\{I_{\Gamma_m}(w_{t_n+s_m} - w_{t_n+s_{m-1}}) | \mathcal{F}_{t_n+s_{m-1}}\}$$

$$= P\{B, (\bar{w}_{s_1}, \bar{w}_{s_2} - \bar{w}_{s_1}, ..., \bar{w}_{s_{m-1}} - \bar{w}_{s_{m-2}}) \in \Gamma_1 \times ... \times \Gamma_{m-1}\}$$

$$\times P(\bar{w}_{s_m} - \bar{w}_{s_{m-1}} \in \Gamma_m)$$

$$= P(B)P\{(\bar{w}_{s_1}, \bar{w}_{s_2} - \bar{w}_{s_1}, ..., \bar{w}_{s_m} - \bar{w}_{s_{m-1}}) \in \Gamma_1 \times ... \times \Gamma_m\}.$$

Therefore, $P(AB) = P(A)P(B)$ for A from a π-system generating

$$\sigma(\bar{w}_{s_1}, \bar{w}_{s_2} - \bar{w}_{s_1}, ..., \bar{w}_{s_m} - \bar{w}_{s_{m-1}}) = \sigma(\bar{w}_{s_1}, \bar{w}_{s_2}, ..., \bar{w}_{s_m}).$$

Since both sides of $P(AB) = P(A)P(B)$ are measures in A, they coincide on this σ-field. Now $P(AB) = P(A)P(B)$ for any A of type

$$\{\omega : (\bar{w}_{s_1}, \bar{w}_{s_2}, ..., \bar{w}_{s_m}) \in \Gamma_1 \times ... \times \Gamma_m\}.$$

The collection of those sets is again a π-system, this time generating $\mathcal{G}_t$. Since both sides of $P(AB) = P(A)P(B)$ are measures, they coincide for all $A \in \mathcal{G}_t$. The lemma is proved.

Lemmas 2 and 3 imply the following.

4. Corollary. *For $t \geq t_n$ and $x \in \mathbb{R}^{d_1}$ the random vector $\xi_t(x)$ and σ-field $\mathcal{F}_{t_n}$ are independent.*

In the following lemma we use the notation $[x] = ([x^1], ..., [x^{d_1}])$ for $x = (x^1, ..., x^{d_1})$.

5. Lemma. *Let $\xi^m = 2^{-m}[2^m \xi_{t_n}]$, where ξ_t is the solution of (2). Then $\xi_t(\xi^m) \xrightarrow{P} \xi_t$ as $m \to \infty$ for each $t \geq t_n$.*

Proof. On the set $\{\omega : \xi^m = x\}$ (a.s.) for $t \geq t_n$ the process $\xi_t(\xi^m)$ satisfies equation (2) with x replaced by ξ^m. Since the union of such sets is Ω, the process $\xi_t(\xi^m)$ satisfies equation (2) (a.s.) with x replaced by ξ^m. We have already noticed above that ξ_t for $t > t_n$ satisfies (2) with ξ_{t_n} in place of x. By noticing that $\xi^m \to \xi_{t_n}$ (uniformly in ω) we get the result as in Theorem 9.3. The lemma is proved.

6. Theorem. *The solution of equation (9.2) is a Markov process.*

Proof. Take $t \geq t_n$ and a bounded continuous function $f(x) \geq 0$. Define

$$\Phi(x) = Ef(\xi_t(x))$$

and let Γ_m be the countable set of all values of $2^{-m}[2^m x], x \in \mathbb{R}^{d_1}$. Since $\xi_t(x)$ is continuous in probability, the function Φ is continuous. Therefore, for $B \in \mathcal{F}_{t_n}$ by Corollary 4 and Lemma 5 we obtain

$$EI_B f(\xi_t) = \lim_{m\to\infty} EI_B f(\xi_t(\xi^m)) = \lim_{m\to\infty} \sum_{r\in\Gamma_m} EI_B f(\xi_t(r)) I_{\xi^m = r}$$

$$= \lim_{m\to\infty} \sum_{r\in\Gamma_m} EI_{B,\xi^m=r}\Phi(r) = \lim_{m\to\infty} EI_B\Phi(\xi^m) = EI_B\Phi(\xi_{t_n}).$$

By definition and properties of conditional expectations this yields (a.s.)

$$E(f(\xi_t)|\mathcal{F}_{t_n}) = \Phi(\xi_{t_n}),$$

$$E(f(\xi_t)|\xi_{t_n}) = E\{E(f(\xi_t)|\mathcal{F}_{t_n})|\xi_{t_n}\} = E(\Phi(\xi_{t_n})|\xi_{t_n}) = \Phi(\xi_{t_n}),$$

$$E(f(\xi_t)|\xi_{t_1},...,\xi_{t_n}) = E(f(\xi_t)|\xi_{t_n}).$$

It remains to extend the last equality to all Borel bounded f. Again fix a $B \in \mathcal{F}_{t_n}$ and consider two measures

$$\mu(\Gamma) = P(B, \xi_t \in \Gamma), \quad \nu(\Gamma) = EI_\Gamma(\xi_t)P(B|\xi_{t_n}).$$

If f is a step function, one easily checks that

$$\int f(x)\,\mu(dx) = EI_B f(\xi_t),$$

$$\int f(x)\,\nu(dx) = Ef(\xi_t)E(I_B|\xi_{t_n}) = EI_B E(f(\xi_t)|\xi_{t_n}).$$

These equalities actually hold for all Borel bounded functions, as one can see upon remembering that such functions are approximated uniformly by step functions. Hence, what we have proved above means that

$$\int f\,\mu(dx) = \int f\,\nu(dx)$$

for all bounded continuous $f \geq 0$. We know that in this case the measures μ and ν coincide. Then the integrals against them also coincide, so that for any Borel bounded f and $B \in \mathcal{F}_{t_n}$ we have

$$EI_B f(\xi_t) = EI_B E(f(\xi_t)|\xi_{t_n}).$$

This yields $E(f(\xi_t)|\mathcal{F}_{t_n}) = E(f(\xi_t)|\xi_{t_n})$. The theorem is proved.

12. Hints to exercises

2.2 If ξ_t is right continuous, then $\xi_t = \lim \xi_{\kappa_n(t)}$, where $\kappa_n(t) = 2^{-n}[2^n t] + 2^{-n}$.

2.4 If $a = -1$ and $b = 1$, then for every $t \geq 0$

$$\{\omega : \tau > t\} = \{\omega : \sup_{r \in \rho \cup \{t\}} \xi_r^2 < 1\},$$

where ρ is the set of all rational numbers on $[0, \infty)$.

2.9 Define $\tau = \inf\{t \geq 0 : |\xi_t| \geq c\}$ and use Chebyshev's inequality.

2.10 In Theorem 2.8 and Exercise 2.9 put $N = c^2$ and integrate with respect to c over $(0, \infty)$.

3.6 Use Exercise 2.9.

3.11 Use Davis's inequality.

3.12 See the proof of Theorem 2.4.1 in order to get that $\tau < 1$ and $(1-s)^{-1} I_{s<\tau} \in \mathcal{S}$. Then prove that, for each $t \geq 0$, on the set $\{t \geq \tau\}$ we have (a.s.)

$$\int_0^t \frac{1}{1-s} I_{s<\tau}\, dw_s = \int_0^\tau \frac{1}{1-s}\, dw_s.$$

3.13 Use Davis's inequality.

4.3 Use Exercise 3.2.5 and Fatou's theorem for conditional expectations.

6.2 In (i) consider $\{\tau \leq t\}$.

6.5 Approximate stopping times with simple ones and use Bachelier's theorem.

7.4 In (iii) use the fact that

$$\int_0^n s^{-1} e^{-|x|^2/(2s)}\, ds - \int_0^n s^{-1} e^{-1/(2s)}\, ds = \int_n^{n|x|^{-2}} s^{-1} e^{-1/(2s)}\, ds \to -2 \ln |x|$$

as $n \to \infty$.

8.4 For appropriate stopping times $\tau_n \to \infty$, the processes $\rho_{t\wedge\tau_n}$ are martingales on $[0, T]$ and the processes $\rho^p_{t\wedge\tau_n}$ are submartingales. By Doob's inequality conclude that $E \sup_{t \leq T} \rho^p_{t\wedge\tau_n} \leq N$.

8.10 Observe that for $\mu = \lambda\tau^{1/2}$ and $r = w_\tau \tau^{-1/2}$ we have

$$\exp(\lambda w_\tau - \lambda^2\tau/2) = \exp(\mu r - \mu^2/2) =: f(r, \mu).$$

Furthermore, Leibniz's rule shows that $f_\mu^{(2k)}(r, 0)$ is a polynomial (called a Hermite polynomial) in r of degree $2k$ with nonzero free coefficient.

10.3 In (i) take any smooth decreasing function $u(x)$ such that $u(x) > 0$ for $x < c$ and $u(x) = 0$ for $x \geq c$, and prove that $u(\xi_t) = \int_0^t u'(\xi_s)b(\xi_s)\,ds$. By comparing the signs of the sides of this formula, conclude that $u(\xi_t) = 0$.

10.4 Observe that $E\xi_{t\wedge\tau(r)} = E(t \wedge \tau(r))$.

Bibliography

[Bi] Billingsley, P., Convergence of probability measures. Second edition. Wiley Series in Probability and Statistics: Probability and Statistics. A Wiley-Interscience Publication. John Wiley & Sons, Inc., New York, 1999.

[Do] Doob, J. L., Stochastic processes. Wiley Classics Library. A Wiley-Interscience Publication. John Wiley & Sons, Inc., New York, 1990.

[Du] Dudley, R.M., Real analysis and probability. Chapman & Hall/CRC, Boca Raton-London-New York-Washington, D.C., 1989.

[GS] Gīhman, Ĭ.Ī.; Skorokhod, A.V., The theory of stochastic processes. I. Grundlehren der Mathematischen Wissenschaften [Fundamental Principles of Mathematical Sciences], 210, Springer-Verlag, Berlin-New York, 1980.

[IW] Ikeda, N.; Watanabe, S., Stochastic differential equations and diffusion processes. North-Holland Publishing Company, Amsterdam-Oxford-New York, 1981.

[It] Itô, K., *On stochastic differential equations*. Mem. Amer. Math. Soc., No. 4 (1951).

[IM] Itô, K., McKean, H.P., Diffusion processes and their sample paths. Grundlehren der Mathematischen Wissenschaften [Fundamental Principles of Mathematical Sciences], 125, Springer-Verlag, Berlin, 1965.

[Kr] Krylov, N.V., Introduction to the theory of diffusion processes, Amer. Math. Soc., Providence, RI, 1995.

[Me] Meyer, P. A., Probability and potentials, Blaisdell Publishing Company, A Division of Ginn and Company, Waltham, Massachusetts, Toronto, London, 1966.

[RY] Revuz, D.; Yor, M., Continuous martingales and Brownian motion. Third edition. Grundlehren der Mathematischen Wissenschaften [Fundamental Principles of Mathematical Sciences], 293. Springer-Verlag, Berlin-New York, 1999.

[Sk] Skorokhod, A. V., Random processes with independent increments. Mathematics and its Applications (Soviet Series), 47. Kluwer Academic Publishers Group, Dordrecht, 1991.

[St] Stroock, D.W., Probability theory, an analytic view (revised edition). Cambridge Univ. Press, 1999.

[SW] Stroock, D.W.; Varadhan, S.R.S., Multidimensional diffusion processes. Springer Verlag, Berlin-New York, 1979.

[Ya] Yaglom, A. M., Correlation theory of stationary and related random functions. Vols. I and II. Springer Series in Statistics. Springer-Verlag, Berlin-New York, 1987.

Index

Selected Titles in This Series

(Continued from the front of this publication)

3 **William W. Adams and Philippe Loustaunau,** An introduction to Gröbner bases, 1994

2 **Jack Graver, Brigitte Servatius, and Herman Servatius,** Combinatorial rigidity, 1993

1 **Ethan Akin,** The general topology of dynamical systems, 1993